T. Bowen, P. Emsley, Jan Ingenhousz, H. Payne

Experiments upon Vegetables

Discovering their Great Power of Purifying the Common Air

T. Bowen, P. Emsley, Jan Ingenhousz, H. Payne

Experiments upon Vegetables
Discovering their Great Power of Purifying the Common Air

ISBN/EAN: 9783337372439

Printed in Europe, USA, Canada, Australia, Japan

Cover: Foto ©berggeist007 / pixelio.de

More available books at **www.hansebooks.com**

EXPERIMENTS

UPON

VEGETABLES,

DISCOVERING

Their great Power of purifying the
Common Air in the Sun-shine,

AND OF

Injuring it in the Shade and at Night.

TO WHICH IS JOINED,

A new Method of examining the accurate
Degree of ~~██████████████████~~

By JOHN INGEN-HOUSZ,

Counsellor of the Court and Body Physician
to their IMPERIAL and ROYAL MAJESTIES.
F. R. S. &c. &c.

LONDON:
Printed for P. ELMSLY, in the Strand;
and H. PAYNE, in Pall Mall. 1779.

Sir JOHN PRINGLE, Bart.

Phyfician to his Britannic Majefty, late Prefident of the Royal Society, Member of the Royal Academy of Paris, &c. &c.

S I R,

A GRATEFUL remembrance of paſt ſervices is as juſt a tribute due to thoſe from whom they are received as the acquitting of a debt contracted in any other manner. If it is not in the power of a man to make a return ſuitable to the benefits received, he is, however, in

duty

duty bound to fhew, by the beft method in his power, a thankful heart to his benefactor.

Ingratitude was by the ancient Greeks held as a crime of the blackeft dye, as tending directly to deftroy the motives of mutual benevolence, and to diffolve the ties of friendfhip, that fource of human happinefs, without which life itfelf is fcarcely worth enjoying.

The ungrateful, confcious of his mifbehaviour, and looking upon his benefactor as upon a judge who has pronounced a juft and fevere fentence upon him, endeavours to find reafon for breaking off with him; while his benefactor, looking upon the ungrateful as upon a monfter unworthy

unworthy of his regard, is induced to fhut, for the future, his heart againft others.

No man upon earth can have ftronger reafons for a due fenfe of gratitude than I acknowledge to you. You beftowed many civilities upon me, who had never been in the way of doing you any fervice whatever. You granted me your friendfhip almoft as foon as I was acquainted with you. You encouraged my eagernefs for improving myfelf in medical knowledge, by communicating to me what you had learned by a laborious life; by that experience which an affiduous and moft attentive zeal, beftowed in the care of the great military hofpitals in

the

the time of war, and a moft fuccefsful private practice, had afforded
you, and of which your celebrated
work upon *The Difeafes of the Army*
will be an honourable and everlaft-
ing teftimony, as well as a real benefit, to the lateft pofterity.

You always gave me, with the
greateft fincerity, your advice in
what manner, and in whofe com-
pany, I could moft improve myfelf
in the various branches of medicine
and natural knowledge which I took
a delight to cultivate.

It was you, SIR, who, among
thofe many refpectable and learned
men, whofe conftant friendfhip
towards me has made an indelible
and grateful impreffion on my mind,

con-

contributed principally to that particular happinefs I enjoyed during fo many years in this ifland; that felicity which a free and independent man finds in the purfuit of knowledge and wifdom in the fociety and friendly intercourfe of thofe who have diftinguifhed themfelves by their learning.

But, SIR, among the many obligations which I owe you, there is one of fuch importance, that the very thought of it ftrikes me with reverence and with the deepeft fenfe of gratitude for you. You did me a fervice which I cannot forbear to mention; though I know that your modefty would hardly permit me to

exprefs

exprefs the true fituation of my
mind in that refpect.

Permit me, SIR, to leave behind
me fome public teftimony of my re-
fpectful gratitude to you, as the
only, though fmall, return I can
make you; the only way by which
I can publicly fhew, that the unfo-
licited favours fo generoufly beftowed upon a foreigner, who could
not claim the leaft merit with you,
have made fo ftrong an impreffion
on my mind as no time is able to
weaken. You have recommended
me, SIR, without my foliciting any
favour from you, to thofe Auguft
Sovereigns who are ftill the fupport
of the illuftrious HOUSE OF AUSTRIA;

thofe

thofe powerful Monarchs whofe gracioufnefs, benevolence, and magnanimity, equal the fupreme grandeur of their ftation. Thefe Auguft Sovereigns, after having fuffered fo many repeated loffes by that dreadful difeafe the Small-pox, refolved at laft to check that terrible havock in their illuftrious Family, and ordered their Ambaffador to fend to their Court a phyfician from this ifland, capable of fulfilling the important truft of faving, by means of inoculation, the remainder of the Royal Offspring, which had as yet efcaped the infection. Being confulted on the choice of a proper perfon, you propofed me without hefitation, and thus opened

[x]

opened to me a wide door to emo-
luments and honours.

After having been fo publicly and
fo honourably called from a diftant
country to the moft generous and
powerful Monarchs; and after having contributed to the tranquillity
and happinefs of fo many illuftrious
Princes, who, being educated under
the maternal care of the moft vir-
tuous PRINCESS, are become highly
important to mankind, and have
filled the world with a well-founded
confidence to fee its happinefs
promoted by their means; whatever advantage or reputation I have
acquired from fuch a flattering ap-

3   pointment,

pointment, I derive it all from your friendſhip.

My earneſt deſire of not quitting this country without leaving you ſome public teſtimony of my real ſentiments towards you, excited me to hurry this work to the preſs without having time enough to finiſh it as I deſired. If it had been in my power to have ſpent the enſuing winter in this country, I might poſſibly have made it more worthy your patronage, and of appearing in the world under your auſpices. I preſent it to you imperfeᴄt as it is; and beg of you to look upon it only as a public mark of my reſpeᴄt and gratitude, which I ſhall retain in full

force

force to the end of my life, and with which I have the honour of fubfcribing myfelf,

S I R,

Your very much obliged

and faithful friend and fervant,

J. INGEN-HOUSZ.

London,
October 12, 1779.

PREFACE.

THE common air, that element in which we live, that invifible fluid which furrounds the whole earth, has never been fo much the object of contemplation as it has in our days: it never engaged fo much the attention of the learned as it has of late years. This fluid, diffufed every where, *the breath of life*, deferves fo much the more the attention and inveftigation of philofophers, as it is the only fubftance without which we can fcarce fubfift alive a fingle moment, and whofe good or bad qualities have the greateft influence upon

our

our conftitution. The moft active poifons which are known do not fo quickly deftroy the life of an animal as the want of air, or the breathing of it when it is rendered highly noxious. It will appear in this work, that thofe very plants, which, influenced by the light of the fun, repair the injury done to this fluid by the breathing of animals, and by many other caufes, may, in different circumftances, poifon fo much this very element, as to render it abfolutely unfit for refpiration, and, inftead of keeping up life, to extinguifh it in a moment. Therefore this univerfally-diffufed element deferves not only the purfuit of philofophers, but claims more immediately the attention of thofe whofe profeffion it is to preferve health and to cure difeafes. I have

beftowed

beſtowed ſome labour upon this
ſubject, both as a philoſopher and
as a phyſician.

When I firſt found in the works
of that excellent philoſopher and
inventive genius, the reverend Dr.
Prieſtley, his important diſcovery,
that plants wonderfully thrive in
putrid air; and that the vegetation
of a plant could correct air fouled
by the burning of a candle, and re-
ſtore it again to its former purity
and fitneſs for ſupporting flame,
and for the reſpiration of animals;
I was ſtruck with admiration: and I
could not read afterwards, but with
a kind of extaſy, the application
which Sir John Pringle made of
this diſcovery in his elaborate diſcourſe, delivered at the Royal Society in November 1773, when he
conferred, as preſident of that
learned

learned Body, the annual prize me-
dal upon Dr. Prieftley, decreed to
him as an honourable teftimony of
their approbation of the fuccefsful
labours beftowed by him upon the
doctrine of air. " From thefe difco-
" veries," fays he, " we are affured,
" that no vegetable grows in vain,
" but that, from the oak of the foreft
" to the grafs in the field; every in-
" dividual plant is ferviceable to
" mankind; if not always diftin-
" guifhed by fome private virtue,
" yet making a part of the whole,
" which cleanfes and purifies our
" atmofphere. In this the fragrant
" rofe and deadly night-fhade co-
" operate: nor is the herbage, nor
" the woods that flourifh in the
" moft remote and unpeopled re-
" gions, unprofitable to us, nor we
" to them; confidering how con-
" ftantly

" ſtantly the winds convey to them
" our vitiated air for our relief, and
" for their nouriſhment. And if
" ever theſe ſalutary gales riſe to
" ſtorms and hurricanes, let us ſtill
" trace and revere the ways of a be-
" neficent Being, who not fortui-
" touſly, but with deſign, not in
" wrath, but in mercy, thus ſhakes
" the waters and the air together, to
" bury in the deep thoſe putrid and
" peſtilential effluvia which the ve-
" getables upon the face of the
" earth had been inſufficient to con-
" ſume." ——

Since I read that elegant diſcourſe,
I have wiſhed that ſome induſtrious
philoſopher would beſtow his labour
in tracing Nature in its operations,
and in diſcovering the manner in
which the vegetable kingdom is
ſubſervient to the animal, in cor-

b   recting

recting the mafs of air contaminated
by their refpiration, and perhaps
too by their perfpiration. The following pages will fhew, whether
the pains I took, in the courfe of this
fummer, to inveftigate this important fubject, have been attended with
fome degree of fuccefs. I am far
from thinking that I have difcovered the whole of this falutary
operation of the vegetable kingdom;
but I cannot but flatter myfelf, that
I have at leaft proceeded a ftep farther than others, and opened a new
path for penetrating deeper into this
myfterious labyrinth.

Among the various ufeful difcoveries with which Dr. Prieftley has
already enriched, and ftill continues
to enrich, natural knowledge, none,
in my opinion, are of more importance

ance than thofe he made upon the various kinds of airs.

The difcovery of that wonderful aerial fluid, which in purity and fitnefs for refpiration fo far exceeds the beft atmofpheric air, that an animal may protract its life in it five times longer than in the beft common air, excites fo much the greater admiration, as he found it firft in fuch bodies which by their nature muft have been fufpected to conceal rather within their fubftance deleterious qualities, fuch as calcined mercury and red precipitate. He has given to this air the very proper appellation of *dephlogifticated air*, or air deprived of that inflammable principle which is the chief ingredient that renders our atmofpheric air more or lefs impure, and thus more or lefs fit for refpiration.

His

His difcovery of that peculiar quality which nitrous air poffeffes, of deftroying or being deftroyed by common air in proportion to its purity, is one of thofe inventions whofe utility will be more and more confpicuous, when it fhall have undergone all the improvements of which it is fufceptible. Let it be mentioned to his honour, that his candour and modefty have made him under-rate the value of this ufeful production of his inquiries, when he fays, in his laft work, intitled, *Experiments and Obfervations relating to various Branches of Natural Philofophy, with a Continuation of the Obfervations of Air,* p. 269, " When I firft difcovered the pro-
" perty of nitrous air as a teft of
" the wholefomenefs of common
" air, I flattered myfelf that it might
" be

" be of confiderable practical ufe;
" and, particularly, that the air of
" diftant places and countries might
" be brought and examined together
" with great eafe and fatisfaction:
" but I own, that hitherto I have
" rather been difappointed in my
" expectation from it." And he
concludes thus: " I have frequently
" taken the open air in the moft
" expofed places in this country, at
" *different times of the year,* and in
" different ftates of the *weather,*
" &c.; but never found the differ-
" ence fo great as the inaccuracy,
" arifing from the method of
" making the trial, might eafily
" amount to or exceed."

Since I faw the manner of putting
different airs to the nitrous teft,
which Abbé Fontana now makes
ufe of, and which I have in my in-

quiries

quiries for the moſt part imitated, I
cannot but think more favourably
of the importance of this diſcovery
than the author himſelf does. I
even think with the Abbé, that, by
uſing convenient and accurate inſtruments, and by obſerving to the
greateſt nicety all the manoeuvres of.
the operation conſtantly in the ſame
way, we may with as much preci-
ſion judge of the degree of purity
of common air, as we now are able
to judge of its degree of heat and
cold by a good thermometer.

Indeed, by this method, even all
the changes which the conſtitution
of the atmoſphere undergoes daily,
in the ſame place, are obſerved with
ſo much accuracy, that, by making
ten obſervations with the ſame air,
the difference will ſcarcely amount

to

to $\frac{1}{500}$th of the two airs employed in the experiment.

The difcovery of Dr. Prieftley, that plants thrive better in foul air than in common and in dephlogifticated air, and that plants have a power of correcting bad air, has thrown a new and important light upon the arrangement of this world. It fhews, even to a demonftration, that the vegetable kingdom is fubfervient to the animal; and, *vice verfâ*, that the air, fpoiled and rendered noxious to animals by their breathing in it, ferves to plants as a kind of nourifhment. But in what manner this faculty of the plants is excited remained ftill unknown.

There was even fome doubt left in the mind of many philofophers, whether the facts related by Dr. Prieftley were not owing to fome

par-

particular accident, as they had by no means been uniform; nay, had even been often contradictory, as he himself candidly owns (fee vol. I. p. 91, &c. of Dr. Prieftley's work on the fubject of air, and his laft work, p. 296.); and as Mr. Sheele had conftantly obferved a contrary effect from beans.

Dr. Prieftley acknowledges, p. 299. that, by repeating (1778) again his experiments, *they proved to be unfavourable to his former hypothefis.* " For," fays he, " whe-
" ther I made the experiments with
" air injured by refpiration, the
" burning of candles, or any other
" phlogiftic procefs, it did not grow
" better but worfe; and the longer
" the plants continued in the air,
" the more phlogifticated it was."
He proceeds thus farther: " I have
" had

" had feveral inftances of the air
" being undoubtedly meliorated by
" this procefs, efpecially by the
" fhoots of ftrawberries, and fome
" other plants, which I could, by
" bending, introduce into the jars
" or phials of air fupported near
" them in the garden, when the
" roots continued in the earth.——
" I had other inftances, no lefs un-
" queftionable, of common air not
" only receiving no injury, but even
" confiderable advantage, from the
" procefs, having been rendered in
" fome meafure dephlogifticated by
" it, fo as to be much more dimi-
" nifhed by nitrous air than before,
" a thing which I was far from ex-
" pecting.——In moft of the cafes in
" which the plants failed to me-
" liorate the air, they were either
" manifeftly fickly, or at leaft did
" not

" not grow and thrive, as they did
" moſt remarkably in my firſt ex-
" periments at Leeds, the reaſon of
" which I cannot diſcover.—In
" thoſe inſtances in which the plants
" grew the beſt, they were, how-
" ever, but ſickly, as appeared by the
" leaves ſoon turning yellow, and
" falling off when the leaſt motion
" was given to them. In ſome
" caſes, however, as in thoſe men-
" tioned in vol. I. p. 91, I ſaw no
" particular reaſon why the air
" ſhould not have been meliorated.
" Upon the whole, I ſtill think
" it *probable*, that the vegetation
" of healthy plants, growing in ſi-
" tuations natural to them, has a
" ſalutary effect on the air in which
" they grow. For one clear in-
" ſtance of the melioration of air in
" theſe circumſtances ſhould weigh
" againſt

" againſt a hundred caſes, in which
" the air is made worſe by it."

Soon after, p. 305, he relates
ſeveral inſtances in which a plant
had, in the ſpace of ſeven, eight,
ten, or more days, effectually mended the foul air in which it was
made to grow. P. 309, he relates
a fact, in which a ſprig of *winter
ſavory*, kept growing in a jar from
the 16th of June to the 20th, had
improved the air evidently, which
improvement he found by three repeated trials to be in the proportion
of 1.275 to 1.375. He relates
another inſtance, in which air was
ſo much improved by a ſprig of
parſley growing in it, from the 16th
of June to the 1ſt of July, that one
meaſure of it with one of nitrous
air occupied only one meaſure.

After

After all, he concludes with the following words, p. 310: " When " thefe obfervations are well confi- " dered, I think it will hardly be " doubted but that there is fome- " thing in the procefs of vegetation, " or at leaft fomething ufually *at-* " *tending* it, that tends to meliorate " the air, in which it is carried on, " whatever be the *proximate caufe* " of this effect, whether it be the " plants imbibing the phlogiftic " matter, as part of their nourifh- " ment, or whether the phlogifton " unites with the vapour that is " continually exhaled from them ; " though of the two opinions I " fhould incline to the former."

Mr. *Sheele* is fo far from thinking that air is meliorated by plants, that he even maintains, that vege-tation has the fame effect on air

that

that refpiration has. He allows, however, that plants do not grow fo well in dephlogifticated as in common air.

At the end of Section XXXIII, in which he treats *of the fpontaneous emiffion of dephlogifticated air from water in certain circumftances*, he fpeaks thus : " It will probably be " imagined, that the refult of the " experiments recited in this Sec- " tion throws fome uncertainty on " the refult of thofe recited in this " volume, from which I have con- " cluded, that air is meliorated by " the vegetation of plants, efpecially " as the water, by which they were " confined, was expofed to the open " air, and the fun, in a garden. " To this I can only fay, that I was " not then aware of the effect of " thefe circumftances, and that I " have

" have reprefented the naked *facts*
" as I obferved them ; and, having
" no great attachment to any par-
" ticular *hypothefis*, I am very will-
" ing that my reader fhould draw
" his own conclufions for himfelf."
Dr. Prieftley, having obferved that
bubbles of air feemed to iffue fpon-
taneoufly from the ftalks and roots
of feveral plants kept in the water,
fufpected immediately, that perhaps
this air, if found better than common air, had been percolated through
the plant, and purified by leaving
its phlogifton in the plant as its
nourifhment. With this view he
plunged many phials containing
fprigs of mint in water, laying
them in fuch a manner, as that
any air, which might be difcharged
from the roots, would be retained
in the phials, the bottoms being
a little

a little elevated. In this pofition
the fprigs of mint grew very well,
and in fome of the phials he obferved a quantity of air to be collected, though very flowly ; but he
was much difappointed, that fome
of the moft vigorous plants produced no air at all. At length,
however, from about ten plants he
collected, in the courfe of a week,
about half an ounce-meafure of air,
which he found fo pure, that one
meafure of it and one of nitrous
air occupied the fpace of only one
meafure.

This remarkable fact contributed
not a little to confirm his faith in
the hypothefis of the purification
of the atmofphere by vegetation ;
but he did not enjoy this fatisfaction long; for, as he found that
other plants of the fame fpecies

I   produced

produced no fuch effect, and that, what he thought more extraordinary, the phials, in which the above mentioned plants had grown, the infide of which were covered with a green kind of matter, continued to yield air as well when the plants were out of them as they had done before ; he was convinced, that the plants had not, as he had imagined, contributed any thing to the production of this pure air. See Dr. Prieftley's laft work, p. 337 and 338.

Thus far this matter was carried on when I took it up in June laft. I muft acknowledge, that, from what is above related from Dr. *Prieftley's* works, I had little doubt but there was fome quality in plants proper for correcting bad air, and improving ordinary air. My curiofity led me to inveftigate in what manner this

this operation is carried on, whether the plants mend air by abforbing, as part of their nourifhment, the phlogiftic matter, and leaving thus the remainder of the air pure (to which opinion Dr. Prieftley inclines the moft) ; or whether perhaps the plants poffefs fome particular virtue hitherto unknown, by which they change bad air into good air, and good into better, which I fufpected to be the cafe.

I was not long engaged in this enquiry before I faw a moft important fcene opened to my view: I obferved, *that plants not only have a faculty to correct bad air in fix or ten days, by growing in it, as the experiments of Dr. Prieftley indicate, but that they perform this important office in a compleat manner in a few hours ; that this wonderful operation*

c   *is*

*is by no means owing to the vegetation
of the plant, but to the influence of
the light of the sun upon the plant.
I found that plants have, moreover, a
most surprizing faculty of elaborating
the air which they contain, and undoubtedly abforb continually from the
common atmosphere, into real and fine
dephlogisticated air; that they pour
down continually, if I may so express
myself, a shower of this depurated air,
which, diffusing itself through the common mass of the atmosphere, contributes to render it more fit for animal
life; that this operation is far from
being carried on constantly, but begins
only after the sun has for some time
made his appearance above the ho-
rizon, and has, by his influence, prepared the plants to begin anew
their beneficial operation upon the air,
and thus upon the animal creation,*

which

which was ftopt during the darkneſs
of the night ; that this operation of
the plants is more or leſs briſk in proportion to the clearneſs of the day,
and the expoſition of the plants more
or leſs adapted to receive the direct
influence of that great luminary ;
that plants ſhaded by high buildings,
or growing under a dark ſhade of
other plants, do not perform this
office, but, on the contrary, throw
out an air hurtful to animals, and
even contaminate the air which ſur-
rounds them ; that this operation of
plants diminiſhes towards the cloſe of
the day, and ceaſes entirely at ſun-ſet,
except in a few plants, which continue
this duty ſomewhat longer than others;
that this office is not performed by
the whole plant, but only by the
leaves and the green ſtalks that ſup-
port them ; that acrid, ill-ſcented,

and

and even the moſt poiſonous plants
perform this office in common with
the mildeſt and the moſt ſalutary;
that the moſt part of leaves pour
out the greateſt quantity of this dephlogiſticated air from their under ſurface,
principally thoſe of lofty trees; that
young leaves, not yet come to their
full perfection, yield dephlogiſticated
air leſs in quantity, and of an inferior quality, than what is produced
by full-grown and old leaves; that
ſome plants elaborate dephlogiſticated
air better than others; that ſome
of the aquatic plants ſeem to excell
in this operation; that all plants
contaminate the ſurrounding air by
night, and even in the day-time
in ſhaded places; that, however,
ſome of thoſe which are inferior to
none in yielding beneficial air in
the ſun-ſhine, ſurpaſs others in the
power

power of infecting the circumambient air in the dark, even to such
a degree, that in a few hours they
render a great body of good air so
noxious, that an animal placed in
it loses its life in a few seconds;
that all flowers render the surrounding air highly noxious, equally by
night and by day; that the roots removed from the ground do the same,
some few, however, excepted; but
that in general fruits have the same
deleterious quality at all times, though
principally in the dark, and many
to such an astonishing degree, that
even some of those fruits which are
the most delicious, as, for instance,
peaches, contaminate so much the
common air as would endanger us to
lose our lives, if we were shut up in a
room in which a great deal of such
fruits are stored up; that the sun by

itself

itſelf has no power to mend air with-
out the concurrence of plants, but on
the contrary is apt to contaminate
it.

Theſe are ſome of the ſecret
operations of plants I diſcovered in
my retirement, of which I will en-
deavour to give ſome account in
the following pages ; ſubmitting,
however, to the judgement of the
candid reader the conſequences,
which I thought might fairly be
deduced from the facts I am to
relate.

I muſt not omit to acquaint the
reader, that, in purſuing the experiments related in this work, he will
find that he labours in vain, if he does
not make uſe of pump-water freſhly drawn ; for if this water has
been expoſed to the open air during
ſome time, it will have parted with

5   a great

a great deal of its own air, and will
therefore be apt to abforb the air
from the leaves. It may alfo happen, that every pump-water may
not be found equally as good as
that which I met with in my country dwelling, though as yet I have
no pofitive reafon to think fo;
but I have fome grounds to believe,
that water drawn from an open well
is far inferior in goodnefs to that
which is forced up by a pump, as
the former is too much expofed to
the open air.

By cafting an eye upon the experiments related in this work, it
will be eafily underftood, why, in
every experiment of this kind,
fome difference in the refult will
commonly be obferved; for the
peculiar degree of goodnefs of the
dephlogifticated air obtained from

the

the leaves depends upon too many circumſtances to be conſtantly of the ſame quality. Some more or leſs light of the ſun thrown upon the jar will make ſome difference; the leaves being more or leſs crowded together, will make a remarkable difference, as a great number of them may be ſhaded from the ſun by others.

As I made the greateſt part of my experiments according to the preſent method of proceeding of my reſpectable friend the Abbé Fontana, it would have been difficult to imitate the experiments related in this work, and even to underſtand the manner in which they were made, if he had not given me leave to anticipate the publication of his own ingenious contrivance, and of his preſent method

thod of putting the different fpecies of air to the teft. This kind-
nefs of that gentleman deferves my
public thanks.

Inaccuracies in the manner of
expreffing myfelf will find fome
indulgence in a man born and educated in the Republic of the United
Provinces, and who was not early
in life acquainted with the Englifh
language.

The AUGUST SOVEREIGNS, whom
I have the honour to ferve, conde-
fcending gracioufly to prolong my
leave of abfence, and allowing me
to fpend the prefent fummer in
this ifland, I thought it my duty to
apply the time granted me by their
goodnefs to an ufeful purpofe, and
to make all the advantage I could of
that peculiar degree of health which
I have

I have always enjoyed in this climate.

On purpofe to avoid every caufe of obftructing my mind in the clofe purfuit of the object I had in view, and in tracing Nature in its operation on this fubject, I difengaged myfelf from the noife of the metropolis, and retired to a fmall villa, where I was out of the way of being interrupted by any body in the contemplaion of Nature.

This work is a part of the refult of above 500 experiments, all which were made in lefs than three months, having begun them in June, and finifhed them in the beginning of September, working from morning till night. From thefe experiments fome more confequences might have been drawn, if I had had more time to employ myfelf

in

in a work upon fuch important matter. Whatever I have been able to deduce from my labours is done in a hafty manner, as my ftay in this country was far too limited to allow me to compofe my work in a regular and more fatisfactory manner.

Though I was very far from forefeeing all the difcoveries which I made in the courfe of this fummer, yet I was perfuaded that a good deal of the oeconomy of the vegetable kingdom might be difcovered by a fteady purfuit of experiments tending to trace the operations of Nature. I had this object in view fome years ago; but, as I did not enjoy fuch a favourable difpofition of mind and body as was neceffary for a tafk, in which all poffible fteadinefs, perfeverance,

verance, and clofe attention were requifite, I deferred the undertaking till I fhould find myfelf more fit for it.

Detached experiments may indeed be very ufeful when a fufficient number is collected to draw fome conclufions from them; but, without purfuing methodically the fame object, difcoveries are to be expected only by mere chance, and are even fometimes overlooked. I owe to the example of my worthy friend, the Abbé Fontana, the thorough perfuafion, which I now entertain, that natural knowledge can make but a very flow progrefs in the hands of thofe who have not patience and affiduity enough in purfuing one and the fame object, till they difcover fome things undifcovered before; or till they find
that

that the difficulty of the undertaking furpaffes their abilities.

WHEN this book was entirely printed, and nothing but the latter end of the preface unfinished, I was informed by my friend the Abbé Fontana, that he difcovered a few days ago a new method of procuring to a fick perfon the benefit of breathing any quantity of dephlogifticated air at a cheap rate.

This very year a paper of mine was read before the Royal Society, and ordered for the prefs (containing a new theory of the effects of gunpowder, and the difcovery of a new and powerful explofive air), in which I fay, that the rapid progrefs our modern philofophers daily

3   make

make on the different kinds of air, induce me to believe that we touch at the happy moment, at which a very eafy and little expenfive method of producing this beneficial fluid, in any quantity wanted, will be produced for the cure of many difeafes.

I have the great fatisfaction to inform the reader, that my prediction is fulfilled even before it hath reached the public, and that this important *defideratum* in medicine is difcovered.

Abbé Fontana found that an animal breathing-in either common or dephlogifticated air renders it unfit for refpiration by communicating to it a confiderable portion of fixed air, which is generated in our body, and thrown out by the lungs as excrementitious. This fixed air is eafily

abforbed

abforbed by fhaking it in common water, but infinitely more readily by the contact with quick-lime water.

He fills one of the large receivers of an air-pump, which are very wide, at their upper extremity, half full of dephlogifticated air extracted from nitre, fo that it may contain about 500 cubic-inches of this air, which will ferve for breathing during half an hour. The manner of drawing this air out of the receiver, is either by thrufting a bended glafs tube under the receiver (when this is floating in water, in which it is fupported by its peculiar bulky form), reaching into the air itfelf, and keeping the other extremity in the mouth; thus drawing this air in the lungs, and breathing it out by the fame tube. This air returning from the

the lungs is infected by fixed air, which being immediately abforbed by the contact with lime-water, the dephlogifticated air is reftored very near to its former purity. Inftead of the bended tube juft mentioned, the top of the receiver may be made as the neck of a bottle, and a tube may be fixed to it, having a cock to open and fhut the paffage as required.

We confume, by each infpiration, about 30 cubic-inches of air; and thus, allowing 15 infpirations for a minute, we confume each minute 450 cubic-inches of air. The Abbé Fontana found, that the dephlogifti-cated air being, after each refpiration, purified again by the lime-water, will remain good about thirty times as long as it would when breathed in the ordinary way; and that

that thus the quantity of dephlogifticated air neceffary for one minute will now ferve for breathing
during half an hour, and thus the
expences will be thirty times lefs.

We confume, in the fpace of 24
hours, about 21600 cubic-inches of
air. One pound of nitre yields by
heat about 12000 cubic-inches of
.dephlogifticated air, and thus it
yields more air than any patient
could confume by breathing this
beneficial air the whole day (for we
muft allow at leaft 12 hours in the
24 for fleoping and neceffary occu-
pations), becaufe this quantity will
ferve thirty times longer by the
method explained, than in the ordinary way. It follows by this,
that the expences required for
breathing a whole day dephlogifti-

cated air cannot amount to one ſhilling.

I have only juſt time enough to announce this happy diſcovery to the publick; whoſe great utility will, I truſt, ſoon be found in the curing of inflammatory and putrid diſeaſes, &c. in which too great a quantity of inflammable principle is let looſe in our blood.

I HAVE

I HAVE alſo diſcovered, ſince my book was printed, that, in reading Dr. Prieſtley's laſt work, I had overlooked a remarkable paſſage, p. 270, in which he hints at what I found to be the cafe with inflammable air having ſtood a long while with plants. I think it my duty to relate his own words: "I am "ſatisfied, however, from my own "obſervations, that air may be "very offenſive to the noſtrils, pro- "bably hurtful to the lungs, and "perhaps alſo in conſequence of "the preſence of phlogiſtic matter "in it, without the phlogiſton be- "ing ſo far *incorporated with it*, "as to be diſcovered by the mix- "ture of nitrous air."

E R R A T A, neceſſary to be corrected.

Page 35, l. 20, before the words *we may find* add *if we could reach the final cauſes of all the operations of nature,*

Page 43, l. 7, for *of which quantity gathered* read *which quantity was gathered*

Page 67, l. 5, for *leaving* read *living*

Page 97, l. 11, after *nitrous* add *and common*

Page 120, l. 5, read *which time may diſcover how to ſeparate, and thus*

Page 142, l. 10, *and this atmoſphere* read *and the air*

Page 187, l. 2, read *by Abbé Fontana's method, becauſe only one meaſure of dephlogiſticated air was employed.*

Page 290, l. 6, read *of bubbles when it was heated in the ſun, was ſo much recovered, that freſh leaves*

AS this volume may possibly fall into the hands of some who have not yet read the works of Dr. Prieftley, and are entirely ftrangers to the new doctrine of air; I think it will be ufeful to explain the meaning of the new invented names given to different kinds of air mentioned in this book.

Nitrous air is that kind of *permanent elaftic fluid* which is extracted by nitrous acid or *aqua fortis* from the moft part of metals, principally from mercury, brafs, copper, &c. This air is by a bended glafs tube conducted under

d 3 an

an inverted jar full of water. The
nitrous air, thus difengaged, rifes
through the water, and fettles at
the inverted bottom of the jar.
Mercury yields the ftrongeft nitrous
air, and always of the fame quality; but heat muft be applied for
its extrication, if a greater quantity
is in a fhort time required. I ufe
for convenience fake copper, from
which nitrous air is extracted in
abundance in a fhort time, without
applying heat. The nitrous acid
muft be diluted with water for the
purpofe.

Inflammable air is that air which
rifes up in bubbles from ftagnant
waters, whofe bottoms are marfhy,
when the ground is ftirred up with
a ftick. It is alfo extracted from
iron, zinc, and fome other metals,
by diluted vitriolic or marine acid.
· This

This air has in common with almoſt all other inflammable ſubſtances, that it is not ſuſceptible of actual inflammation, without it be in contact with common or reſpirable air. This air has the following qualities by which it may be known: it is not abſorbed by ſhaking it in water; it is not diminiſhed by the addition of nitrous air; it is inſtantly and abſolutely mortal to animals breathing in it; it burns by the approach of the flame of a candle, where it is in contact with common air; but the whole of it inflames at once, and explodes with a loud report, when it is intimately mixed with common, and principally with dephlogiſticated air.

Phlogiſticated air is, properly, air impregnated with phlogiſton, or with the inflammable principle. It

d 4   has

has received this name becaufe
common air, expofed to the calcination of metals, becomes phlogifticated air; which alteration feems to
depend on the phlogifton of the
metal being communicated to it, for
the metal itfelf has loft it in the
calcination; and becaufe the burning of a candle, and many other
phlogiftic proceffes, impart to common air this quality. The air returning from our lungs is faid to
be phlogifticated more or lefs, becaufe it is found to poffefs the fame
qualities as the air expofed to the
calcination of metals. This kind
of air is known by the following
properties: it is not abforbed by
water; it is not, or not much, diminifhed by nitrous air; it is poifonous
for an animal who breathes in it;
it is not inflammable either by it-

felf

felf or by the addition of refpirable air, but extinguifhes flame.

Dephlogifticated air is that pure,
etherial, permanent, and elaftic fluid
which the celebrated Dr. Prieftley
found out the firft, and gave this
very well adapted name to it. It is
refpirable air, deftitute of the phlogiftic or inflammable principle with
which the beft atmofpheric air is
found always to be more or lefs contaminated. It is in reality common air of the higheft purity, fuch
as never exifts in the common atmofphere. It is even fo far fuperior
in purity to common air, that an
animal fhut up in a veffel, filled
with this air, protracts its life four
or five times, nay even in fome cafes
feven times longer, than if it was
fhut up in a veffel filled with the
beft atmofpheric or common air.

Some

Some of its qualities, by which it may be known, are the following : it is not abforbable by the contact of water; the flame of a candle plunged in it becomes larger, and of the moft admirable brilliancy, fo as to dazzle the eyes; red-hot charcoal plunged in it becomes fhining and fparkling; it is much more diminifhed by nitrous air than common air; it explodes, with an uncommon loud report, when mixed with a certain proportion of inflammable air, or when a few drops of good vitriolic æther are poured in a veffel containing this air, as I difcovered.

Fixed air is that kind of aerial fluid which iffues in abundance from fermenting fubftances, and which, in fome places, rifes out of the ground by itfelf, as in the fa-
mous

mous *Grotta del Cane* near Naples. It is this air with which fome mineral waters are impregnated, and to which they owe their pungent tafte and their virtue, as, for inftance, Seltzer waters. It is that air which arifes in abundance from calcarious ftones, by the addition of vitriolic acid. This air may be known by the following properties: it extinguifhes flame; it is abforbed by water, and communicates to it the fame pungent, acidulous tafte as Seltzer water has, fo as not to be diftinguifhed from it either by the tafte or by its virtues; it precipitates quick lime from water; it immediately cryftallizes *oleum tartari per deliquium,* when put in a veffel lined with this oil; it is mortal to animals breathing in it.

Eudio-

Eudiometer, is a new word; it ſignifies an inſtrument by which we may judge of the degree of ſalubrity of the common air. The invention of ſuch an inſtrument belongs to Dr. Prieſtley. It conſiſts chiefly of a glaſs tube, divided in equal parts; for inſtance, in two large diviſions; each of which is divided into ten others, and each of theſe ten ſub-divided again into ten parts: and a glaſs meaſure, containing exactly one of the great diviſions of the tube. One meaſure of common air and one of nitrous air, put together in a ſeparate glaſs veſſel, and left by themſelves till the diminution of the bulk of the two airs is compleated, and afterwards let up in the glaſs tube, indicates at once the exact diminution of the two joint meaſures. The degree of

6   goodneſs

goodnefs of the common air is found to be in proportion to the diminution of the bulk of the two airs. Mr. Magellan, F. R. S. has publifhed a work upon an ingenious contrivance of his own of this kind, which is to be fold by Mr. Parker in Fleet-Street, with the proper directions how to ufe it. What confiderable improvements we owe in this regard to Abbé Fontana, will appear in the introduction to the fecond part of this work.

ADVER-

ADVERTISEMENT.

AS the author intends to publiſh
a French tranſlation of this
work, he thinks it his duty to
give public notice of his in-
tentions, that no one may give
himſelf any unneceſſary pains
about it.

CON-

CONTENTS.

Sect.

CONTENTS. lxv

e PART

Sect.

ON

O N   T H E

N A T U R E

O F

P L A N T S.

S E C T I O N I.

Some general remarks on the nature
of the leaves of plants, and their
ufe.

IT feems to be more than probable, that the leaves, with which
the moft part of plants are fur-
niflhed during the fummer in tem-
perate climates, and perpetually in

hot countries, are deſtined to more
than one purpoſe. Such a great
apparatus, which nature diſplays as
ſoon as the ſun begins to afford a
certain degree of warmth upon the
ſurface of the earth, can ſcarcely be
conſidered as ſolely deſtined either
to ornament, to nouriſhment of the
plant, to its growth, to ripen its
fruit, or for any other peculiar
and ſingle uſe. It ſeems probable, that they are uſeful to the
growth of the tree; for, by depriv-
ing the tree of all its leaves, it is
in danger of decay. By taking a
conſiderable part of the leaves from
a fruit tree, the fruit is leſs perfeᴄt;
and by taking them all away, the
fruit decays and falls before its ma-
turity. It is alſo probable, that the
tree receives ſome advantage from
the leaves abſorbing, by their means,
moiſture

moifture from the air, from rain, and from dew; for it has been found a confiderable advantage to the growth of a tree, to water the ftem and the leaves now and then. But I leave the difcuffion of thofe articles to others, who have made thefe confiderations an object of their purfuits. The late Mr. Baker has publifhed to the world his microfcopical obfervations on the fubject. Mr. Bonnet, of Geneva, has publifhed a very elaborate work upon the fame, intitled, *Recherches fur l'ufage des Fueilles dans les Plantes, et fur quelques autres Sujets relatif à l'Hiftoire de la Vegetation, par* Charles Bonnet, *à* Gottingen *et* Leiden, 1754. This work contains a great deal of interefting inquiries upon the nature, properties, and utilities of thofe wonderful organs; all which

have

have been treated with the greateſt attention, and have thrown much light upon this ſubject.

This celebrated author has taken a great deal of notice of thoſe air bubbles which cover the leaves when plunged under water. He ſays, p. 26, that the leaves draw theſe bubbles from the water. He is the more perſuaded that this is the caſe, becauſe he found theſe bubbles did not appear when the water had been boiled ſome time, and appeared more when the water is impregnated with air, by blowing in it. He had alſo obſerved, that they did not appear after ſunſet. Page 31, he explains his opinion farther upon this head: he ſays, that theſe air bubbles are produced by common air adhering to the external ſurface of the leaves,

3   which

which fwells up into bubbles by the heat of the fun; and that the cold of the night is the reafon why thefe air bubbles do not make their appearance at that time. As he found that dry leaves put under water gather fuch bubbles alfo upon their furface, he concludes, p. 33, that the appearance of thefe bubbles cannot be owing to any vital action in the leaves.

I took fome pains to difclofe the caufe of thefe bubbles, which, I think, are of more importance than Mr. Bonnet at that time imagined them to be; and found the fact to be generally this:

The moft part of leaves gather thefe bubbles upon their furface, when plunged in any water in the fun-fhine or by day-time in the open air; but infinitely more in

frefh

frefh pump water than in any other.
In clear river water they appear
later, lefs in number and in fize;
lefs fo in rain water, and the leaft
of all in boiled water, in ftagnating,
and in diftilled water.

They are not produced by the
warmth of the fun rarifying the
air adhering to the leaves; for many
kinds of leaves produce them almoft
as foon as plunged under water,
though the water be very cold, and
the leaves warm from the fun-fhine
be plunged in it.

They do not appear after fun-fet,
at leaft not in any confiderable number; but thofe that already exift do
not fhrink in or difappear by the
cold of the night.

As foon as the fun begins to dif-
fufe its warmth over the furface of
the earth in the fpring, and to promote

mote that general tendency to corruption which all dead bodies of the
animal and vegetable kingdom, and
many other ſubſtances, are ſo liable
to, the trees diſplay in a few days
the moſt wonderful ſcene that can
be imagined. Contracted as they
were in that ſtate of ſtupor and in-
activity in which they remain during
the winter, expoſing to the air no
other ſurface than that of their
trunk and branches, as if they
wanted to have as little to do as poſſible with the external air, they all
at once ~~inſtantly, perhaps more~~ than
a thouſand times, their ſurface by
diſplaying thoſe kind of numberleſs
fans which we call leaves. Some
of them produce their leaves a long
while before any flowers appear
upon them; others a good while
after the flowers are formed, and

the fructification is already in an advanced ftate; and keep their leaves in the beft condition, and even pufh out continually new ones, long after the whole fructification is finifhed; which feems to indicate, that the chief ufe of thefe fans is not to affift the fructification and propagation of their fpecies. Thefe fans, when compleated, feem to compofe or arrange themfelves in fuch a manner as to expofe their upper and varnifhed furface to the direct influence of the fun, and to hide as much as they can their under furface from the direct influence of this luminary. It feems as if they required rather the light of the fun than the influence of its heat, as their polifhed furfaces muft reflect fome of the rays of the fun, and thus moderate the degree of heat.

It

It will, perhaps, appear probable, that one of the great laboratories of nature for cleanfing and purifying the air of our atmofphere is placed in the fubftance of the leaves, and put in action by the influence of the light; and that the air thus purified, but in this ftate grown ufelefs or noxious to the plant, is thrown out for the greateft part by the excretory ducts, placed chiefly, at leaft in far the moft part of plants, on the undet fide of the leaf.

Is there not fome probability that the under part of the leaves may have been chiefly deftined for this purpofe, becaufe in this way the dephlogifticated air, guſhing continually out of this furface, is inclined to fall rather downwards, as a beneficial fhower for the ufe of the animals who all breathe in a region of the

the air inferior to the leaves of trees? Does not this conjecture get some weight, if we consider that dephlogisticated air is in reality specifically heavier than common air, and thus tends rather to fall downwards?

If we add to these reflexions another of no less importance, *viz.* that most sorts of foul air are specifically lighter than common air, we shall be inclined to believe that the difference of the specifical gravity of that beneficial air of which I treat, and that which is become hurtful to our constitution by corruption, breathing, and other causes, indicates one of those special blessings designed by the hand of God: for by this arrangement we get soon rid, in a great measure, of that air which is become hurtful to us, as it rises soon up out of our reach; whereas the

the dephlogifticated air, being heavier than common air, is rather inclined to fettle on the furface of the earth among the animal creation.

But, as animals fpoil equally as much air in the winter as in the fummer by the act of refpiration, it might feem fomewhat furprizing, that this great laboratory ceafes intirely by the decay of the leaves. Is this defect fupplied by fome other means equally powerful? Though we are very far from being able to trace all the active caufes which contribute their fhare in keeping up the wholefomenefs of our atmofphere, yet we have already traced fome of them, and therefore muft not defpair of difcovering fome more. The fhaking of foul air in water will in great meafure correct it. Water itfelf has a power of

of yielding dephlogisticated air, as Dr.
Priestley discovered. Plants have a
power to correct bad air, and to improve good air. Winds will blow
away the noxious particles of the
air, and bring on air corrected by the
waters of the seas, lakes, rivers,
and forests. All these causes exist
equally in the winter as in the summer, or at least nearly so. The influence of the vegetable creation
alone ceases in the winter: but the
loss of this influence is, perhaps, more
than amply counterbalanced by the
diminution of the general promoting
cause of corruption, viz. heat. Every
body knows, that warm weather
hastens in a great degree putrefaction. In the summer time number-
less insects are produced, which did
not exist in the winter: these insects
infect the air by the corruption of
their

their bodies. That immenfe quantity of animal fubftances, and many others, which undergo a putrefaction by the warmth of the weather, feems to require an additional power or agent to counter-act it; and this office is deftined to the leaves. In frofty weather no animal fubftance is fubject to putrefaction, which cannot go on without a proper degree of heat. The perfpiration of animals is lefs offenfive in the winter than in the fummer, and of confequence muft corrupt the atmofphere lefs. It feems therefore probable, that, if we are deprived of one way by which air is corrected in the winter, we have alfo at that time lefs caufes which tend to contaminate our element.

S E C-

SECTION II.

*On the manner in which the dephlo-
gifticated air is obtained from the
leaves of plants.*

AS the leaves of plants yield dephlogifticated air only in the clear
day-light, or in the fun-fhine, and
begin their operation only after they
have been in a certain manner prepared, by the influence of the fame
light, for beginning it; they are to
be put in a very tranfparent glafs
veffel, or jar, filled with frefh pump
water (which feems the moft adapted
to promote this operation of the
leaves, or at leaft not to obftruct it);
which, being inverted in a tub full
of the fame water, is to be immediately expofed to the open air, or
rather

r

rather to the fun-fhine: thus the leaves continuing to live, continue alfo to perform the office they performed out of the water, as far as the water does not obftruct it. The water prevents only new atmofpheric air being abforbed by the leaves, but does not prevent that air, which already exifted in the leaves, from oozing out. This air, prepared in the leaves by the influence of the light of the fun, appears foon upon the furface of the leaves in different forms, moft generally in the form of round bubbles, which, increafing gradually in fize, and detaching themfelves from the leaves, rife up and fettle at the inverted bottom of the jar: they are fucceeded by new bubbles, till the leaves, not being in the way of fupplying themfelves with new atmofpheric

fpheric air, become exhaufted. This air, gathered in this manner, is really dephlogifticated air, of a more or lefs good quality, according to the nature of the plant from which the leaves are taken, and the clearnefs of the day-light to which they were expofed.

It is not very rare to fee thefe bubbles fo quickly fucceeding one another, that they rife from the fame fpot almoft in a continual ftream: I faw this more than once, principally in the *nymphæa alba*.

SEC-

SECTION III.

*The manner in which the dephlogifti-
cated air oozes out of the leaves
is different in different plants.*

IT is fomewhat amufing to obferve the conftant manner in which
the dephlogifticated air makes its
appearance upon the fame kind of
leaves, and the furprizing difference
which takes place in the leaves of
different plants. Some leaves, for inftance, form always fmall round bubbles, as is the cafe with the moft part
of leaves; others form irregular flat
blifters, as do the leaves of the honey-
fuckle, *caprifolium.* Some, and indeed the greateft part, produce round
bubbles on both furfaces; others
yield on one furface round bubbles,

on

on the other irregular blifters; for inftance, leaves of oak, which give flat blifters on the under fide, and round bubbles on the upper **fide.** Some form only thofe irregular blifters at the upper fide, **as the leaves** of fpurge, *cataputia* or *euphorbia*.

Some leaves **form neither** bubbles nor **blifters** on either **fide, and yet** yield **a great deal of dephlogifticated** air; for inftance, leaves of *nafturtium Indicum:* thefe leaves feem to have a quality repulfive to water, **which** only forms a kind of cover over the **furface of the leaves, without com-** ing **into immediate** **contact.** The air oozing out of the leaves **floats** under this cover, and rifes to the higheft part of the leaves, where it forms a kind of bags, which at laft detach themfelves from the **edge,** and rife to the top of the jar. The
leaves

leaves themfelves, after ftanding a
day and a half in the water, are not
wetted by it, but come out quite
dry.

Some leaves have that peculiar
quality of being wetted by the
water only on one fide; as, for inftance, leaves of rafberry fhrubs,
which do not receive the wet on
their fhaggy under furface.

Strawberry leaves repel the water
from both furfaces, form blifters at
the under furface, and chiefly round
bubbles at the upper furface.

Some leaves begin very early in
the morning to yield dephlogifticated air, and ceafe late in the evening; for inftance, potatoe and *malva*
leaves.

Some begin this operation very
late in the morning, and ceafe very

early

early in the evening; for inftance, leaves of *laurocerafus*.

Some leaves yield the air bubbles immediately, as leaves of potatoe plants; fome in a few feconds, as *malva*; fome in a few minutes, as walnut trees; fome much later, as leaves of *laurocerafus*.

Some yield the air bubbles firft on the under fide, as almoft all leaves of trees; fome firft at the upper fide, as leaves of *laurocerafus*; fome on both furfaces at the fame time, as *malva*.

On fome leaves the air bubbles grow almoft all regularly one with another, as in leaves of a vine, walnut, lime tree; in fome they are, from the very beginning, of a very irregular fize, as in *malva*, parfley, &c.

Thefe

These few inſtances ſhew the various ways in which this beneficial air is oozing from the leaves, and which can only be owing to the different organization of the leaves in different plants.

I have diſcovered a good deal more of ſuch remarkable peculiarities among leaves of different plants; but thoſe above mentioned will be ſufficient to ſhew, that each plant follows in this regard its own nature; and that, therefore, theſe different appearances ſeem to depend upon ſome vital motion going on in the ſubſtance of the leaves.

SEC-

SECTION IV.

The dephlogisticated air oozing out of the leaves in the water is not air from the water itself.

THE reverend Dr. Prieſtley found, that water, chiefly pump water, ſtanding ſome days by itſelf, forms at the bottom and ſides of the veſſel a kind of green matter, ſeemingly vegetable, from which air bubbles riſe continually to the top of the jar, if expoſed to the ſun-ſhine: that this air is fine dephlogiſticated air, which ſhews that there is a faculty in water to produce by itſelf this beneficial fluid; and thus, that the maſs of the waters of the ſea, lakes, and rivers, have their ſhare in purifying the atmoſphere.

But

But as this dephlogifticated air is not produced immediately from the pump water, but only when this green matter is formed, it is clear, that the air obtained from the leaves, as foon as they are put in the water, is by no means air from the water, but air continuing to be produced by a fpecial operation carried on in a living leaf expofed to the day-light, and forming bubbles, becaufe the furrounding water prevents this air from being diffufed through the atmofphere.

It is true, that pump water, placed in the fun-fhine, will foon yield fome fmall air bubbles, fettling at the bottom of the jar, and every where at the fides; but this air is very far from being the fame as that contained in the air bubbles of the leaves.

I placed,

I placed, in a warm fun-fhine, a great number of inverted jars, full of pump water, and collected care-fully from them all the air yielded by thefe bubbles, which proved to be much worfe than the common air.

I boiled fome pump water in a pot, in which I had placed a long cylindrical jar, quite full of the fame water: a good deal of air was col-lected at the top of the inverted jar, which was by the heat difengaged from the water. This air proved to be much worfe than common air, and entirely unfit for refpiration.

Abbé Fontana has made, fome years ago, a great many experi-ments, tending to inveftigate the nature of air contained in different waters.

S E C-

SECTION V.

The dephlogisticated air oozing out of the leaves in the water is not existing in the substance of the leaves in this pure state, but is only secreted out of the leaves when it has undergone a purification, or a kind of transmutation.

IF the dephlogisticated air collected from the leaves in the sun existed in them in its pure state, it must appear as such when squeezed out of the leaves under water; or, at least, if the leaves are only shook gently under water, without hurting their organization, or when they are put in warm or in boiling water.

I squeezed a handful or two of potatoe leaves under water, and kept

an

an inverted jar full of water above it, to receive the air. A great deal of it was inſtantly obtained, which proved to be nearly as good as common air.

I ſqueezed, in the ſame way, ſome air out of leaves of ſage, *ſalvia*, which proved to be ſomewhat worſe than the former.

A potatoe plant was ſhook under water, ſo as not to hurt it: a good deal of air was immediately diſengaged, which, by the nitrous teſt, proved to be worſe than common air.

A plant of *lamium album* was treated in the ſame way, and in like manner a good deal of air was obtained, which was nearly of the ſame quality with the former.

Some leaves of an apple tree were put in a cylindrical jar full of pump water. The jar was then inverted

in

in a veffel full of the fame water, and placed upon the fire. As foon as the water grew warm, the leaves were covered with air bubbles, juft as in the fun. After the water had boiled a little while, it was put by to cool. A great deal of air was obtained, which proved to be fo bad as to extinguifh flame.

Some of the fame leaves were put into a jar, inverted in a pot full of water, and only placed near the fire: a great deal of air was obtained, but as poifonous as the former.

SEC-

SECTION VI.

*The production of the dephlogisticated
air from the leaves is not owing to
the warmth of the sun, but chiefly,
if not only, to the light.*

IF the sun caused this air to ooze
out of the leaves by rarifying the
air in heating the water, it would
follow that, if a leaf, warmed in
the middle of the sun-shine upon
the tree, was immediately placed in
water drawn directly from the
pump, and thus being very cold,
the air bubbles would not appear
till, at least, some degree of warmth
was communicated to the water; but
quite the contrary happens. The
leaves taken from trees or plants in
the midst of a warm day, and plunged

I   im-

immediately into cold water, are remarkably quick in forming air bubbles, and yielding the beſt dephlogiſticated air.

If it was the warmth of the ſun, and not its light, that produced this operation, it would follow, that, by warming the water near the fire about as much as it would have been in the ſun, this very air would be produced; but this is far from being the caſe.

I placed ſome leaves in pump water, inverted the jar, and kept it as near the fire as was required to receive a moderate warmth, near as much as a ſimilar jar, filled with leaves of the ſame plant, and placed in the open air, at the ſame time received from the ſun. The reſult was, that the air obtained by the

fire

fire was very bad, and that obtained in the fun was dephlogifticated air.

A jar full of walnut tree leaves was placed under the fhade of other plants, and near a wall, fo that no rays of the fun could reach it. It ftood there the whole day, fo that the water in the jar had received there about the fame degree of warmth as the furrounding air (the thermometer being then at 76°); the air obtained was worfe than common air, whereas the air obtained from other jars kept in the fun-fhine during fuch a little time that the water had by no means received a degree of warmth approaching that of the atmofphere, was fine de-phlogifticated air.

No dephlogifticated air is obtained in a warm room, if the fun does not fhine upon the jar containing the leaves.

S E C-

SECTION VII.

Reflections.

IT might, perhaps, be objected, that the leaves of the plants are never in a natural ftate when furrounded by pump water; and that thus there may, perhaps, remain fome degree of doubt, whether the fame operation of the leaves in their natural fituation takes place. · · ·

I cannot confider the plants kept thus under water to be in a fituation fo contrary to their nature as to derange their ufual operation. Water, even more than they want, is not hurtful to plants, if it is not applied too confiderable a time. The water only cuts off the communication with the external air; and we know,

that

that plants may live a long while without this free communication. Besides, water plants, as *perſicaria urens*, *becabunga*, and others, which I have employed in my experiments, are often found a long while quite covered by the water in which they grow.

By bending a living plant (the root remaining in its own earth) in an inverted jar full of water, you only ſurprize nature upon the fact in the middle of its operation, by ſhutting at once all communication with the free air. In ſuch a ſituation no air can be abſorbed by the leaves, or by any parts of the plant under water; but any air may freely come out of it.

Without covering the leaves or the plant entirely with water, it is impoſſible to know what quantity of of air oozes out of the plant, and of what quality this air is; for any air iffuing out of a plant incorporates immediately with the furrounding air, and makes a compound whofe conftituent parts are an intimate mixture of air from the plant and common air; and it would be as difficult to judge accurately how much dephlogifticated air fuch a plant has communicated to the ordinary air which was already in the jar, as it would be for a chymift to judge accurately what quantity of diftilled water was mixed with a certain quantity of common water, if fome of it was really added to it on purpofe to puzzle him. It may, however, be afcertained, in an inaccurate way, what quantity of this beneficial air a plant, placed in a jar full of common air, has communi-

cated

cated to it, by computing the degree of fuperior goodnefs the air is found to poffefs.

As plants yield in a few hours fuch a confiderable quantity of dephlogifticated air, though their fituation feems rather unfavourable for it when they are kept under water ; may it not with fome degree of probability be conjectured, that they yield much more of it when remaining in ther natural fituation; for then, being continually fupplied by new common air, their ftock of dephlogifticated air cannot be exhaufted. It is an unfavourable circumftance, that air is not an object of our fight; if it was, we fhould perhaps fee that plants have a kind of refpiration as animals have; that leaves are the organs of it; that, perhaps, they have pores

I   which

which abforb air, and others which
throw it out by way of excretion,
as are the excretory ducts of animals; that the air fecreted, being
dephlogifticated air, is thrown out
as noxious to the plant (which article is clearly demonftrated by Dr.
Prieftley and Mr. Sheele); that in
the moft part of plants, principally
trees, the greateft part of inhaling
pores are placed upon the upper fide
of the leaf, and the excretory ducts
principally on the under fide.

If thefe conjectures were well
grounded, it would throw a great
deal of new light upon the arrangement of the different parts of the
globe, and the harmony between
all its parts would become more confpicuous. We might find, that
partial tempefts and hurricanes, by
fhaking the air and the waters, pro-

duce

duce fome partial evils for the uni-
verfal benefit of nature; that, by
thefe powerful agitations, the feptic
and noxious particles of the air are
blown away, and rendered of no
effect, by being thus diluted with
the body of air, and partly buried
in the waters. We might conceive
a little more of the deep defigns
of the Supreme Wifdom in the
different arrangement of fublu-
nary beings. The ftubborn atheift
would, perhaps, find reafon to hu-
miliate himfelf before that Almigh-
ty Being, whofe exiftence he denies
becaufe his limited fenfes reprefent
to him nothing but a confufed chaos
of miferies and diforders in this
world.

SEC-

SECTION VIII.

Dry plants have very little or no power to affect air; but, when moistened, they infect air.

I FILLED a jar with dry hay, another with dry straw, and left it inverted a good while; but could not find the air altered.

I put some leaves of a lime tree, dried for the purpose, in a jar full of fresh pump water; and placed another jar, filled with an equal number of fresh leaves of the same tree, near it in the sun.

The dry leaves began much later than the others to yield round bubbles just in the same way, but which did not grow so quick, nor so large. However, in a few hours, a good

D 3 deal

deal of air was obtained, but of such a bad quality as to extinguish a flame, whereas the fresh leaves had yielded dephlogisticated air: which experiment seems to indicate, that the generation of the dephlogisticated air is owing to the action of the living plant. The same result was also obtained from dry hay put into a jar full of pump water.

SECTION IX.

All plants possess a power of correcting, in a few hours, foul air unfit for respiration; but only in clear day light, or in the sun shine.

THIS remarkable property of plants is indeed very great; for in a
few

few hours, nay even fometimes in an
hour and an half, they purify fo
much a body of air quite unfit for
refpiration, as to be equal in goodnefs to atmofpheric air. They will
even do it when they are inclofed in
a glafs veffel, without any water.
One leaf of a vine, fhut up in an
ounce phial, full of air fouled by
breathing fo that a candle could not
burn in it, reftored this air to the
goodnefs of common air in the
fpace of an hour and a half. But
plants enjoy this privilege only in
the day-time, and when they grow
in unfhaded places.

This power of plants extends itfelf
even to the worft of all airs, in
which an animal finds his deftruction in a moment; fuch as is pure
inflammable and highly phlogifticated air, which is little or fcarcely at

all

all diminishable by nitrous air. I observed some difference in various
kinds of plants in this respect, and
found that water plants seem to
possess this quality in a greater degree than others. The willow tree
and the *persicaria urens* were found
eminent in producing this effect:
and may it not be providentially or-
dained it should be so, as those plants
grow better in marshy, low grounds,
and even in stagnated waters, whose
bottoms are generally muddy, and
yield a great deal of inflammable air,
which may be collected at the surface of the water by stirring up the
ground, and may be kindled by
throwing a burning paper upon the
water, which is an amusing experiment by night? Plants, however,
want longer time to correct this kind
of air, at least that which is extracted

tracted from metals by vitriolic acid.

The property of plants is demon-strated in experiments 41, 51, 56, 57, 58, 59.

SECTION X.

All plants yield a more or less quantity of dephlogisticated air in the day-time, when growing in the open air, and free from dark shade.

THE quantity of dephlogisticated air, and even the quality of it, which the leaves of plants give, seems to be different in different plants: though, indeed, t is may depend in a great measure upon some particular

ticular circumſtances, to which it is not eaſy to be ſufficiently attentive. It ſeems, however, to be a general rule, that the leaves of all plants, growing in a place where they are not much ſhaded by other plants, buildings, &c. yield, in a clear day, dephlogiſticated air; and that this air is yielded in greater abundance, and of a greater purity, when they grow in open places unincumbered by other plants higher than they are themſelves.

I got in general a large quantity of air of a very good quality from ſome water plants, as from the *perſi- cana urens* and willow. The fir trees yielded alſo very fine air, and in abundance.

The *naſturtium Indicum* ſurpaſſed them all in general, in regard as well to the quantity as to the qua- lity.

lity. One hundred leaves of this plant, which are very thin, yielded, in two hours time, as much dephlogifticated air as would fill a cylindrical glafs four inches and a half deep, and one inch and three quarters diameter; of which quantity gathered again afterwards from the fame leaves, without taking them out of the water, fee exp. 30—35. This quantity furpaffes by far the bulk of the leaves themfelves, and fhews to how amazing a quantity the air may amount yielded in a fair day by a lofty tree.

The leaves being more or lefs crouded together, being expofed for a longer or fhorter time, or fooner or later in the day, will occafion fome difference in the quality and quantity of this air.

It

It feems that, in general, the fineft air is obtained when the fun has paffed the meridian.

SECTION XI.

The faculty which plants poffefs of yielding dcphlogifticated air, of correcting foul air, and improving ordinary air, is not owing to the act of vegetation.

IF this wonderful faculty of plants depended upon their vegetation, they would exert it at all times, and in all places in which the vegetation goes on. A plant may vegetate, and even thrive very well, in the utmoft darknefs; and yet in fuch a place

place it has no power to correct bad
air, or to yield good; but, on the
contrary, it ſpreads round about it
deleterious exhalations, which render the beſt air even pernicious to
the utmoſt degree.   .

It will not be difficult to underſtand now from what cauſe all thoſe
different and contrary effects which
Dr. Prieſtley has found in his experiments did really depend; and why
Mr. Sheele had conſtantly found
that the vegetation of beans always
ſpoilt good air.

Theſe gentlemen expected the
good effects from the vegetation of
the plants, as ſuch.   By making a
plant grow night and day in ordinary air kept in a phial with the
plant, the effect will depend upon
the greater or leſs expoſure of the
plant to the light. Beſides, by keeping

ing a plant a long while in pump water, the green matter, of which Dr. Prieftley found to iffue very fine dephlogifticated air, will be generated; and thus the air within the phial, being mixed with this good air, will not in reality indicate the effect of the plant upon this air, as Dr. Prieftley makes no fcruple to acknowledge in his late work, p. 338.

S E C-

SECTION XII.

*The plants evaporate by night bad air,
and foul the common air which
furrounds them; yet this is far
over-balanced by their beneficial
operation in the day.*

THE bad air which plants yield
by night is fo inconfiderable in com-
parifon of the quantity of dephlogifticated air which they yield by
the day-time, that it amounts to
very little. By a rough calculation
I found, that the poifonous air
yielded during the whole night by
any plant could not amount to one-
hundredth part of the dephlogifticated air which the fame plant
yielded in two hours time in a fair
day. But, from my experiments,
one

one might naturally wonder that no remarkable mifchief happens from fo many plants as a foreft contains, if one plant, containing fcarce a handful, may poifon to fuch a degree the quantity of two pints of common air in one night, as to render it abfolutely mortal for any animal who breathes in it.

I make no doubt but a great quantity of plants, kept in a clofe and fmall place during a night, or by day in the dark, may do fome material mifchief, and even occafion death, to any perfon who fhould be imprudent enough to remain in fuch a place. The undoubted facts of people being found dead in their beds, when they had flept in a fmall room with a great deal of flowers in it, muft infpire us with a caution againft keeping too many flowers in close clofe rooms. My experiments go much further, and will, I hope, in future, make people aware of danger if they ftore up a great deal of fruit in a clofe room in which they fleep.

I think that nobody before me even fufpected the leaft danger of keeping beans, peaches, or fuch like fruits, in their rooms; and yet a fufficient number of them would eafily poifon an unwary perfon fleeping in a clofe room, in which an abundance of thefe fruits is ftored.

The gardeners by opening a hot-houfe early in the morning, which has been fhut clofe during the night, or at any time in the day if the fun has not fhined a good deal upon it, are very well aware of a particular oppreffion they feel by entering it. I remember to have felt it more

E   than

than once, without even ſuſpecting
the cauſe of it. Dr. Prieſtley obſerved this remarkable offenſiveneſs
of the hot-houſes with a more phi-
loſophical attention; he tried the air
within them, and found it worſe
than common air.

By all this it ſeems evident, that
it would not be ſafe to ſleep in a
cloſe hot-houſe; that it would not
be prudent to keep too many green
branches, fruit, or flowers, in any
· room by night, particularly in that
of a ſick perſon.

The beſt phyſicians have, indeed,
often recommended to put green
branches of lime-trees and others in
the rooms of their patients, without
ever ſuſpecting any other effect but
benefit from them. I think ſtill,
that ſome benefit may ariſe from
putting, in a clear day, freſh green
branches

branches in the room of a ſick perſon, by cooling the air, which is
owing to the evaporation of moiſture; but I ſhould now apprehend
rather ſome miſchief from them in
a room whoſe doors and windows
are ſhut, and which is not well
lighted. At any rate, I ſhould no
more allow them to be kept in the
night-time in the room of any of
my own patients.

Is it not ſomewhat probable, that
among thoſe people who are found
dead in their bed without any pre-
vious illneſs, ſome may owe their un-
timely end to ſome ſuch concealed
cauſe, which nobody ever ſuſpected
to be in any way dangerous?

But the miſchief which trees in
reality do by night-time to the ſurrounding air, cannot do any ob-
ſerveable harm to animals: for

thoſe

thofe mifchievous exhalations being, very providentially, fpecifically
lighter than common air, rife at
the fame time up; and thus the
lower region, in which we breathe,
is freed from them almoft as foon as
they are produced; whereas the dephlogifticated air iffuing out of the
plants in great abundance in the
day-time is fpecifically heavier than
common air, and is therefore inclined to remain longer among us,
and to afford us all the benefit for
which the Supreme Wifdom has
providentially deftined it.

S E C-

SECTION XIII.

All roots, few excepted, when left out of the ground, yield by day and by night foul air, and infect the surrounding air.

THE experiments I made upon this fubject convinced me that roots have this deleterious power, and fome even to fo great a degree, that it would not be fafe to remain in a fmall clofe room in which a great quantity of roots of plants are kept. The roots of fome aquatic plants are remarkably apt to foul the air in which they are placed, fuch as roots of rufhes, though ever fo well cleaned from mud and dirt, and the roots of *perficaria urens.* But I found the roots of *becabunga*

E 3 almoft

almoſt inoffenſive in this reſpeƈt, which I was the leſs ſurprized at, as their ſubſtance is but little different from the ſtalks. The roots of a muſ-tard plant gave in the ſun-ſhine a good deal of air, when kept under water; which air was worſe than com-mon air, and extinguiſhed a candle. **Theſe roots even corrupted good air in the middle of the ſun-ſhine.**

A plant, with roots and all, placed in a jar full of water, did, however, yield dephlogiſticated air; ſo that the bad effeƈt of the roots was over-balanced by the plant itſelf.

SECTION XIV.

Flowers ooze out by day and night an unwholesome air, and spoil, at any time and in every place, a considerable body of air in which they are placed.

DR. Prieftley has obferved, that a rofe, kept in a glafs, had, in a fhort time, fo much infected the air as to be unfit for refpiration, and concluded from this very juftly, that flowers might be hurtful in a room. I have heard more than once of a perfon found dead in a room where a great deal of flowers were kept; and I make no doubt but fome of thefe cafes are well founded.

I tried all the flowers I could find in my neighbourhood, but could

E 4   not

not difcover one which did not yield
poifonous air, though in a fmall
quantity, by day and by night, and
which had not the power of rendering quite unfit for refpiration a very
confiderable body of common air.
They even feem not to lofe in the
leaft their deleterious influence in
the fun; fo that I cannot but think
that it is unfafe to keep in a clofe
room a large quantity of any
flowers, even fuch as have the moft
delightful fmell. I am, however,
very far from thinking that there
is any danger to apprehend from
fuch nofegays as are commonly kept
in a room, either for ornament or
perfume. The malignant influence
which could be expected from fuch
a fmall quantity of flowers is intirely diffipated in the mafs of the
furrounding air; but the excefs
                                muft

muſt not go too far, if the room is
cloſely ſhut and but ſmall. If a few
flowers of the honey ſuckle (*caprifolium*), which poſſeſs the moſt agreeable ſmell, are able to foul in three
hours time, in the middle of the
day, a body of air equal to two pints
(ſee experiments 65—70) we may
judge what dangerous effect might
be expected from a large quantity
in a cloſe room. Thoſe flowers,
like all others, after having rendered
truly fatal a body of air, have loſt
nothing of their flavour. The air
itſelf, which they have poiſoned, is
impregnated with the ſame fragrant
ſmell as the flowers themſelves; ſo
that a perſon, not aware of the concealed poiſon which flowers ſpread
round about them, might be eaſily
induced by the ſweetneſs of their
ſcent

scent to run the greatest hazard of losing his life, without the smallest apprehension of danger.

SECTION XV.

All fruits in general exhale a deleterious air by day and by night, in the light and in the dark, and possess a remarkable power of spreading a poisonous quality through the surrounding air.

I WAS, indeed, not a little surprized to find this effect in even the most delicious fruit, such as peaches, grapes, apples, and mulberries. By what I observed in my experiments I am apt to think, that the power of
fruit,

fruit, at leaſt of ſome, ſurpaſſes the deleterious quality of flowers in the dark; but the influence of the ſun ſeems to check, in ſome degree, this hurtful quality in ſome fruit more than in roots and flowers, of which the moſt part preſerve, even in the brighteſt ſun-ſhine, their virulent effects upon the ambient air in its full force.   I found, that one peach was able to render intirely poiſonous, in a few hours, a body of air at leaſt ſix times greater than the ſpace it occupied; and even that they could, in the middle of the ſun-ſhine, render ſuch a quantity of air ſo unwholeſome, that a candle could not burn, nor an animal breathe in it.

After I had obſerved, that all leaves of plants yield dephlogiſticated air by day-light; and conſidering, that in general all leaves are green,

green, and that that fubftance which
Dr. Prieftley difcovered to yield
fo much dephlogifticated air is of
the fame colour; I had fome fufpicion, that green fruits, fuch as beans,
would alfo yield dephlogifticated air.
I placed, for this purpofe, fome
French beans in a jar full of pump
water, and expofed it to a bright
fun-fhine during four hours, and
obtained a moderate quantity of air,
oozing out of their fubftance by
bubbles, in the fame manner as
out of leaves. This air was far
from being dephlogifticated air; for
it was even worfe than common air,
though it approached pretty near it
in goodnefs.

I then wanted to fee what effect
thefe green fruits would have upon
a body of air in the dark; and I
was not a little furprized to find,

I   that

that they had a very remarkable power to foul a great body of air, in which they were fhut up, to fuch a degree, that two dozen of fmall French beans, placed in a jar holding two pints, had rendered in one night the air in the jar abfolutely poifonous, fo that a young chicken placed in it was killed in lefs than twenty feconds. I found even this deleterious influence of beans upon air to furpafs the power of plants, which are known to be of a poifonous quality. See experiments 75—91, and principally experiments 88 and 89.

Ripe mulberries, filling one-third of a jar, and placed in the fun during four hours, had fo much fouled the air within the jar that a candle would not burn in it.

SEC-

SECTION XVI.

The power of plants in correcting bad air is greater than their faculty of improving good air.

THE experiments already known of Dr. Prieftley, by which it appears that plants thrive wonderfully well in air fouled by breathing and burning of candles, gave me a great fufpicion, that the power of plants in correcting bad air might furpafs their faculty of improving good common air. In order to put my conjecture to the trial, I placed at eleven o'clock, in a warm fun-fhine, two jars of an equal fize, each containing an equal quantity of fprigs of pepper-mint, in pump water. In one of thefe jars was let up a certain quantity

tity of common air, whofe teft was at that time fuch, that one meafure of it with one of nitrous air occupied 1.06½. In the other jar was let up the fame quantity of air fouled by refpiration, of which one meafure with one meafure of nitrous air occupied 1.34.

The air of both jars was examined at two o'clock, when I found the common air fo much improved, that one meafure of it with one of nitrous air occupied 1.00. The foul air was fo much mended, that it was near as good as the atmofpheric air, for one meafure of it with one of nitrous air occupied now 1.08½.

I examined both airs again at four o'clock, when the common air was ftill more improved; for one meafure of it with one of nitrous air

air occupied 0.95. The foul air now was not only become as good as refpirable air, or air of the atmofphere, but even furpaffed it in goodnefs, for one meafure of it with one of nitrous air occupied 1.05.

Now, as the fame plant brought the foul air from 1.34 to 1.05, and the common air from 1.06½ to 0.95, it appears clear, that the plant had corrected the foul air far more than it had improved the common air.

This experiment was repeated feveral times with nearly the fame refults.

As plants feem to delight in foul air, probably becaufe this air impregnated with phlogifton affords more proper nourifhment, *viz.* phlogifton to the plant; it muft of courfe happen, that a plant draws to it fo much the more phlogifton

as

as the air, in which it grows, contains more of this principle.

When a plant grows in the open
air, it contaminates by night the
furrounding air; but this air, being
diluted with other air, does not appear in reality to be altered by any
method yet found out: befides, it
is probable, that this air is rifen up
as foon as it was become phlogifticated by the plant, being fpecifically
lighter than common air. It feems
therefore not improbable, that fome
plants, as for instance the hyofcyamus,
may contaminate in reality more air
at night than they improve in the day;
fo that, if all the air fpoiled by fuch
a plant was fhut up with the plant
a whole night and a day, the air
would'ftill be found contaminated:
but tho' this might be the cafe when
the plant is fhut up with the air,

F   yet

yet it could never be any real dif-
advantage in the natural fituation
of things, becaufe this fouled air
may be corrected in the atmofphere
by fome manner or other unknown
to us ; and, if not, we are, at any
rate, immediately out of its reach,
as it rifes by its being become lighter. But if fuch infectious plants
are fhut up in fmall clofe rooms,
they certainly might do a material
injury to our conftitution, and even
occafion death.

It appears, by experiment 41,
that a plant may really foul fo much
air at night as fcarce to be able to
correct in the day. But it is to be
confidered, that fuch a plant, being
maimed by its roots being taken off,
and by being fhut up in a narrow
fpace, muft have loft fome of that
vigour which plants naturally have
when

when they remain undiſturbed upon their place. See alſo experiments 51, 56, 57, 58, 59, 60.

SECTION XVII.

*On the effeƈt of leaving plants kept
in a room.*

THO' I think, that the keeping
of a few plants in rooms is very
indifferent as to the health of the
perſons who live in them; yet it
is not ſo indifferent for us to know
the effects which plants have in
reality on the air of the room, that
we may avoid danger from any
exceſs.

F 2   The

The influence of plants on the air of a room in which they are kept is different in the night from what it is in the day. In the day plants are apt to contribute somewhat to purify the air of the room, if they are placed fo as to receive all the light of the fun poffible: if they are placed fo as not to receive the direct influence of the fun, but to be free from any fhade, they feem to have no influence at all, either in improving the air of the room or in fouling it. But when they are placed in a part of the room the moft remote from the windows; fo as to be much fhaded, they are apt to render the air of the room more or lefs impure, in proportion to their fize, and to the more or lefs degree of light of the place where they ftand. At night they

they abfolutely tend to foul the air, principally when they flower. I acknowledge readily, that a few flower-pots can do neither good nor harm. But I remember to have found feveral orange-trees in a room, by way of ornament, and, as I was told, to keep the air of the room wholefome : I think now fuch ornamental plants by no means indifferent, unlefs they were but fmall and the room ample; at any rate I fhould not fuffer them to be kept in a room at night, where a fick perfon is.

A plant fhut up in a glafs jar, and placed near the window in a room fo as to receive the rays of the fun, will make the air of the jar better than the air of the room : whereas a fimilar plant, placed in the fame room in a fhaded place,

F 3   will

will render the air of the jar worfe than the air of the room. If, after a few hours, you invert the experiment, by placing the plant which ftood at the window in the fhade, and that which ftood in the fhade near the window, the reverfe will take place, *viz.* the air of the jar, which was improved, will be found worfe than the air of the room; and the air of the jar, which had been contaminated, will be found corrected again. And this remarkable property of plants, in the way juft mentioned, may be demonftrated in a few hours. See experiment 45.

S E C-

SECTION XVIII.

Leaves of plants die sooner when the dephlogisticated air, elaborated by them, is separated from them.

WHEN the dephlogisticated air, settled in the form of bubbles upon the leaves, is shook off, new bubbles succeed; and thus by shaking off several times these air bubbles a greater quantity of dephlogisticated air is obtained. The second crop of bubbles contains in general a finer kind of air than the first; the reason of which may be, that it is scarce possible to free the surface of the leaves entirely from all atmospheric air sticking to them, particularly those which have a rough

F 4   shaggy

shaggy surface ; as, for instance, the leaves of sage, *salvia*.

Some of these leaves are so prolific in pushing out these bubbles, that I have found them reproduced nine or ten times in leaves of a pear-tree. The leaves of a vine are also very ready to yield a good number of successions of these bubbles. But I was curious to see whether the leaves decay sooner or later when the air bubbles were left upon them, or when they were shook off now and then : I put a leaf of a vine in a jar full of pump-water, and left it exposed to the open air without ever stirring it : the air bubbles grew to a very large size ; and some of them quitted the leaf of themselves and rose up. This leaf remained as fresh as when it was put in the jar during a whole week ;

week; whereas another leaf of the same vine, placed near it in another jar, and whose air bubbles were shook off five or six times in a day, was withered in less than two days. This second leaf had lost the greatest part of the rough surface, which covers, as a kind of scarf-skin, the under and unvarnished part of the leaf; at least this scarf-skin became transparent, if it was not really destroyed; and this transparency was observed principally upon the very spots of the air bubbles. This experiment was repeated several times with the same success.

It should seem by this observation, that the loss of this air, if it cannot be replaced by the absorption of new air from the atmosphere, makes the leaves decay sooner; and thus the texture of
the

the leaves, having no more air to elaborate, refembles almoft the organization of an animal, which lofes its life by becoming exhaufted through the loffes fuftained by the increafe of the various excretions which are carried on in its body, if thefe loffes are not repaired by taking in new nourifhment.

Vegetables feem to draw the moft part of their juices from the earth, by their fpreading roots ; and their phlogiftic matter chiefly from the atmofphere, from which they abforb the air as it exifts. They elaborate this air in the fubftance of their leaves, feparating from it what is wanted for their own nourifhment, *viz.* the phlogifton, and throwing out the remainder, thus deprived of its inflammable principle, as an excrementitious fluid, and in this

ftate

ſtate hurtful to them, but rendered uſeful to the animals, who in their tour take from this air, by the act of reſpiration, what they want, and throw out the remainder as hurtful to them ; but rendered again ſerviceable to the vegetables. This theory ſeems to be very reaſonable, and to have ſome foundation in nature. It throws a good deal of light upon the œconomy of nature, and the mutual influence which the vegetable kingdom has upon the animal, and the animal upon the vegetable. It has ſome analogy with other general operations of nature, which are well known.

A plant, which is a living being, deſtitute of motion, remaining upon the ſame ſpot on which it took its beginning, if not capable, as animals are, of going in ſearch of its food, muſt

muſt find within the narrow
compaſs of the ſpace it occupies
every thing which is wanted for itſelf, and to fulfil the office which
has been dictated to it by the Author of nature. It is obliged to
ſpread the numberleſs filaments of
its root through the ſurrounding
ground, as ſo many ſiphons to pump
up the juice, which preſents itſelf
to thoſe filaments ; and theſe filaments are ſufficient to afford all
that the greateſt part of trees want
in the winter. But, being deſtined
in the ſummer-time to more important offices, the tree ſpreads through
the air thoſe numberleſs fans, diſ-
poſing them, in the moſt advan-
tageous manner imaginable, to in-
cumber each other as little as poſſible in pumping from the ſurrounding air all that they can ab-
ſorb

forb from it, and to prefent, if
I may fo fpeak, this fubftance
drawn from the common atmofphere to the direct rays of the fun,
on purpofe to receive the benefit
which the influence of that great
luminary can give it.

SECTION XIX.

On the power which vegetables have
of abforbing different kinds of air.

IF a plant is fhut up in a certain quantity of air, and all light
hindered from falling upon it, it ab-
forbs in general more air than it
yields, and therefore the bulk of air

is

is found lefs. The quantity of air thus abforbed by plants may vary from numberlefs circumftances, as well as from the particular nature of the plant. I have no time to fearch in my notes for all the particularities I have obferved upon this fubject. I can fay in general from remembrance, that fome water-plants were very willing to abforb a good deal of air, principally when they were placed with roots and all in the air; and that they readily abforbed air fouled by breathing.

One of thofe plants had alfo abforbed a great deal of dephlogifticated air, fo that in one night it had abforbed half the quantity I had put with it, which amounted to 4 ounce meafures, if I rightly remember.

This

This abforption also takes place in the day time; but as the plants at that time yield themfelves a great quantity of air, the abforption is not fo eafily afcertained.

SECTION XX.

On the beft manner of judging whether the plants are ready to yield their dephlogifticated air.

AS the light of the fun, and not the warmth, is the chief caufe, if not the only one, which makes the plants yield their dephlogifticated air, it feems reafonable to think, that in a bright fun-fhiny morning

the

the plants will be earlier revived in their office than when the fun is hid by thick clouds. I found this difference to be very remarkable, fo that in a dark cloudy morning I found the plants to begin their daily operation an hour or two later than in a clearer day. I even found that all the plants in the fame garden did not awake, if I may fo exprefs myfelf, at the fame time from their nocturnal ftupor. Thofe plants, whofe expofure was fuch as favoured the rays of the fun being caft early upon them, were revived earlier than thofe which were fhaded by other plants, a wall, a houfe, &c. Nay, I even found that there was fome difference in this refpect between the leaves of the fame tree; as I found thofe which were the firft influenced by

the

the light of the fun, the firſt ready to operate; when thoſe of the oppoſite ſide of the tree, ſhaded by the firſt, were ſtill in their ſtate of ſtupor.

A ready way to know this time exactly is to put a leaf or two, from the plant you are to examine in this reſpect, in a glaſs full of freſh pump-water, and to obſerve, whether the bubbles appear upon them about as quick as they uſe to do in the full day time. If they do, you may be ſure they are fit for the buſineſs.

But there is a readier way to know exactly this article of time, which I found by the water in the jar in which the green matter, diſcovered by Dr. Prieſtley, is already formed. The doctor found that this green matter yielded air bubbles

only

only when placed in the fun; which obfervation ferved me as a good index, whether thofe plants which have experienced nearly the fame influence of the fun as the green·matter, were fit to yield dephlegifticated air. The more brifk you fee thefe bubbles rife, the quicker your plants will give theirs. But this manner of judging can only be of fervice in the morning; for in the middle of the day all leaves of plants, even thofe which were kept in a very dark place, revive fo quickly, that they feem not to ftand fhort in an obfervable way with thofe which were conftantly in the open air.

SEC-

SECTION XXI.

*Conjectures why some waters, as dis-
tilled, boiled, and some other waters, do not promote, but impede
the operation of the plants in yielding dephlogisticated air.*

AS I think I have proved clearly enough that the dephlogisticated
air yielded by plants is air elaborated
by a kind of vital motion, carried
on in the substance of the leaves,
and kept up by the influence of the
light of the sun, it seems that no
more is required to collect this air
than to prevent its diffusing itself
through the common mass of the
atmosphere. Water seems the most
appropriated body for such an in-
tention, for it is not hurtful to plants.

Many

Many of them even thrive the beſt in it. The beſt quality required therefore in the water uſed for this purpoſe ſeems to be, to poſſeſs of itſelf air enough, ſo as not to imbibe it readily from the plants; and not ſo much as to be overcharged with it; for if the water is too much deprived of its own air, it muſt be more diſpoſed to abſorbe it from bodies plunged into it. And if water ſhould be ſo much impregnated with any air, this air would readily ruſh into the ſubſtance of the leaves, and ſpoil by its bulk, or by its particular nature, the elaboration of the dephlogiſticated air; the more ſo, as water, when found ſaturated with air, is found to poſſeſs this air in the form of fixed air, which differs too much from the nature of dephlogiſticated air,

or

or atmoſpheric air. Beſides, water overcharged with air parts eaſily with it, which of conſequence will of itſelf ſettle in the form of bubbles upon the leaves, and thus diſturb their whole operation. We know that pump-water poſſeſſes of itſelf a great portion of air, which is generally thought to be for a part fixed air, to which it owes its agreeably pungent or briſk taſte, which makes it palatable above all other waters. We know with more certainty, that boiled and diſtilled water are deprived of the greateſt part of their air; and this is perhaps the reaſon, why they are not ſo palatable as common ſpring or pump-water. Therefore it ſeems to be not quite improbable, that water which has been boiled or diſtilled is very apt to abſorbe itſelf the air

which

which oozes out of the leaves, and that thus lefs air is gathered at the top of the bottle. This conjecture will perhaps find more ground from the following experiment. I placed fome leaves of a vine in water, which I had, for this experiment, impregnated with fixed air: they were fcarce under the furface of this water, but they were all covered with air bubbles; which feems to me to depend partly upon this water refufing to abforb any air iffuing from the leaves, becaufe it was already overcharged with air itfelf. It is true that any other body, plunged in water impregnated with fixed air, will alfo become covered with air bubbles; but thefe bubbles do not appear fo foon, or increafe fo rapidly, as thofe of the living leaves. So that it feems, that the bubbles of

the

the leaves increaſe faſter becauſe
they are puſhed out of the leaves by
a vital motion in the leaf. It is alſo
true, that leaves thus placed in water impregnated with fixed air, do
not yield that fine dephlogiſticated
air which they yield when placed
in common pump-water; which
may be owing perhaps to the great
abundance of fixed air penetrating
the leaves, by being abſorbed, and
oozing out as it were, in a kind of
tumultuary way, together with the
air already contained in the leaves.
Thus the air iſſuing out of the
leaves may not have undergone
that degree of elaboration required
to change it into dephlogiſticated
air: for the leaſt circumſtance may
diſturb nature in this work; the
ſhade of a building, or of another
plant, may change this wonderful
G 4 opera-

operation, ſo as to produce quite the reverſe, and to obtain a poiſonous air inſtead of dephlogiſticated air : for the evaporation of bad air in the dark depends on the vital motion within the plant, which, being not influenced by the light of the ſun, produces a contrary effect. Thus a plant growing in an abſolute darkneſs is without green colour, and fruit without the influence of the light has no flavour.

S E C-

SECTION XXII.

Some remarks on the green matter which settles at the bottom and fides of the jars in which water is left standing.

THIS green matter, which feems to be of the vegetable kind, was firſt found by the Rev. Dr. Prieſtley to yield very pure dephlogiſticated air: but it ceaſes at laſt to yield more air if the water of the jar is not renewed, which ought therefore to be done now and then.

It is wonderful that this matter ſeems to be never exhauſted of yielding dephlogiſticated air, though it has no free communication with the common atmoſphere, from which

5   the

the moſt part of other plants ſeem
to derive their ſtock of air. Does
this vegetable matter imbibe the air
from the water, and change it into
dephlogiſticated air? This does not
ſeem to me probable, for I could not
obtain from water, even by boiling,
ſo much air as the water in which
this ſubſtance was produced yielded
by itſelf. I ſhould rather incline to
believe, that that wonderful power
of nature, of changing one ſubſtance into another, and of promoting perpetually that tranſmutation
of ſubſtances, which we may obſerve every where, is carried on in
this green vegetable matter in a
more ample and conſpicuous way.
The water itſelf, or ſome ſubſtance
in the water, is, as I think, changed into this vegetation, and undergoes, by the influence of the ſun
ſhining

shining upon it, in this very substance or kind of plants, such a *me-
tamorphosis* as to become what we
call now dephlogisticated air. This
real transmutation, though wonderful to the eye of a philosopher, yet
is no more extraordinary than the
change of grass and other vegetables into fat within the body of a
graminivorous animal, and the production of oil from the watery juice
of an olive tree. More examples
are to be seen of such wonderful
transmutations of sublunary beings
in the article upon the mutability
of air.

On purpose to obtain in a short
time a great deal of dephlogisticated
air from this green matter, I gathered a good deal of it from the
sides of a stone trough placed near
a spring upon the high road, and
always

always kept full of water for the horfes. I put a good deal of this fubftance in a jar holding a gallon of pump-water, and inverted it in an earthen pan. In a week's time I found about $1\frac{1}{2}$ pint of very fine dephlogifticated air collected in the jar, which furpaffed in purity the air obtained in another jar from the green matter ·generated by itfelf. See experiment 100.

SECTION XXIII.

In planting trees for rendering the air wholesomer, it seems not to be quite indifferent what kind of trees are made use of.

AFTER what is already said on the subject, there will be no doubt left, that vegetables have a remarkable share in cleansing and purifying our atmosphere. But as it seems to follow from my experiments, that some trees yield by the day a purer dephlogisticated air than others, and that some seem to be less disposed to infect common air by night, it can scarce be considered as a matter entirely indifferent what kind of trees ought to be planted,

if

if the falubrity of the air was the chief object of fuch a plantation. I made fome experiments for this purpofe, of which a few are placed in the fecond part of this book. But I am far from thinking myfelf intitled to decide any thing upon this head; the more, becaufe all trees co-operate to the fame end, and becaufe the œconomical advantage arifing from the preference of one fort of tree above another may be thought to over-balance the fmall advantage to be derived from its fuperiority in rendering the air purer. I muft content myfelf with the difcovery of the fact, and leave the reft to others, who, by farther and more decifive experiments, may have a better right to decide fomething upon this head than I can as yet pretend to.

SEC-

SECTION XXIV.

The largest and the more perfect leaves yield more and purer dephlogisticated air, than those which are not yet full grown.

NOTHING feems to me a more convincing proof that the elaboration of dephlogisticated air is an effect of a kind of vital motion in the texture of the leaves, than that young leaves, not yet grown to their natural fize, yield their air-bubbles flower and lefs in bulk, and that the air yielded by full-grown leaves furpaffes in purity that which is obtained from leaves not yet come to perfection.

I

It

It is an amuſing ſight to obſerve in a jar full of pump-water the extremity of a branch of a vine, which contains leaves of different ages, from the matureſt to thoſe which only begin to unfold themſelves. The air-bubbles make firſt their appearance upon the old leaves, then upon thoſe that follow, and laſt of all on the new-born ones. The ſame proportion takes place alſo in the ſize of the bubbles; the largeſt or oldeſt leaves having always the largeſt bubbles, and therefore yielding far the greateſt quantity of dephlogiſticated air.

As it ſeems to be almoſt a conſtant rule, that the leaves which yield the greateſt quantity of air, yield alſo the pureſt; the ſame rule alſo takes place in the old and new leaves. The young leaves
ſeem

feem not to have their organization compleated for the office to which they are deftined, and therefore they are not yet able to elaborate fo much nor fo good dephlogifticated air as the old ones. The experiment 122 and 123 feem to be decifive in this refpect.

SECTION XXV.

Though the diminution of the bulk of nitrous air is believed to be an unqueftionable teft of the goodnefs of any air, yet, it muft be allowed, that in fome kinds of airs this teft may fail.

AFTER having tried a great variety of airs myfelf, and after

H   having

having feen many more tried by
Abbé Fontana, I no longer made the
leaft doubt, but the difcovery of
Dr. Prieftley in judging of the exact degree of goodnefs of any air
was without any exception. But,
as I was refolved to abftain as much,
as poffible from all analogical conclufions, without they were fupported by direct experiments, I tried
every air I could find, not only by
the nitrous teft, but alfo by the
flame of a candle, without, however, harbouring any miftruft in the
already adopted manner of examining the degree of goodnefs of them.

I had already been convinced,
that inflammable air was made explofive in a few hours when expofed
in the fun with any plant, though
I fometimes found it, by the nitrous
teft, fo much corrected as to approach

proach near to the goodnefs of common air. This gave me fome fufpicion, that this inflammable air might be fufceptible of a ftill more remarkable correction or purification, at leaft in appearance, without lofing its explofive quality.

On purpofe to difcover the whole, I left fome inflammable air upon *perficaria*, and fome upon wallnut leaves, during forty-eight hours, keeping the jars continually in the open air.

I tried firft the air of the jar in which the wallnut leaves were, in the manner familiar to Dr. Prieftley and in that of A. Fontana ; and repeated each trial twice with the fame refult. I found the air by both thefe methods to exhibit all the appearance of air, fuperior in quality to common air; as may be feen in

experi-

experiments 110, 111. 113, 114, and 115; and yet I found this very air to explode with fuch a loud report, even in a cylindrical jar, that my fervant, who kept the glafs in his hands, thought it was abfolutely broken. This event gave me no fmall concern for a method of trying the goodnefs of airs, which I had already confidered as infallible. However, I had ftill fome hopes left that I had committed fome blunder in this experiment; and very luckily I had ftill at hand the jar which contained the *perficaria urens* with the inflammable air; but I was forry to find that my fufpicion was but too well grounded; for this air gave, by two different trials, the following refult: one meafure of it with one of nitrous air occupied 0.95; and with two

meafures

meafures of nitrous air 1.92; by A. Fontana's method it gave 1.90, 1.96, 2.95; and thus it did appear by thefe trials to furpafs far in goodnefs the common air; and yet it exploded at the flame of a candle with an uncommon loud report. See experiments 110 and 111.

There remained ftill one experiment to be tried with this air, *viz.* to put a living animal in it. I was forry to have fpent the moft part of this air, fo as not to have enough of it left for this trial. However, I was refolved to pufh the experiment farther, and to let the inflammable air ftand a longer while upon the plants, before it was to be employed for the different trials, and principally before an animal fhould be put in it. Some entire plants of *perficaria urens* were put

in

in a gallon jar full of water, and as much ſtrong inflammable air was let up as to fill above one third of the jar. I left it in the garden during ſix days, when I found, to my ſurprize, that it was very far from being corre_ted; for one meaſure of it, with one of nitrous air, occupied 1.80; it gave the following reſult by Abbé Fontana's method, 2.58, 3.58 : a chicken, near three weeks old, died in it in the ſpace of one minute.

This reſult, ſo different from the former, greatly puzzled me, and reſtored my hope that the nitrous teſt was without exception, and that I muſt have committed ſome error in the former experiment.

I was, however, far from giving up entirely my ſuſpicion of the failure of the nitrous teſt. I re-

ſolved

folved to repeat the experiment a-
gain, with all poffible attention; I
had ftill half a pint left of the inflammable air, which had been during fix days upon the *perficaria urens*
without being much mended. See
exp. 112. I put a frefh plant of
muftard in a jar filled with water, and
let up this inflammable air in the jar,
fo that the plant was in contact with
the air. I placed it in the garden
on a Saturday at twelve o'clock. I
tried this air the next day between
one and two in the afternoon, and
found it by the nitrous teft fo much
mended, that it appeared better than
common air, and yet it exploded
with a loud report by the approach
of a candle. I replaced the jar again
in the garden, and put the fame air
again to the nitrous teft on the Monday following, when it appeared to

H 4   be

be far fuperior to the atmofperic air, for one meafure of it, with one of nitrous air, occupied o.,6; and yet it exploded as ftrongly as before. I replaced it again in the garden during four hours more, when it appeared to be ftill farther improved by the nitrous teft, without lofing, however, in the leaft, its explofive nature. See experiment 115.

I had alfo on the fame Saturday put fome plants of *perficaria urens* with their roots in a jar full of water, and let up two pints of ftrong in-flammable air. I found this air on Sunday, after the jar had been 24 hours in the garden, fo much cor-rected, that it approached to the goodnefs of common air by the ni-trous teft, though it exploded with a loud report. I replaced the jar again in the garden, and again ex-

3   amined

amined the air on Monday between
one and two o'clock, when it appeared, by the nitrous teſt, about as
good as common air ; and yet it
had not loſt its exploſive quality.
After this, I replaced the jar as before, in the garden, and put the
ſame air again to the teſt between
four and five in the afternoon of
the ſame day, when it appeared to
be better than common air, without
having loſt its exploſive force.
There remained now nothing more
to be done, than to try the effect of
this air upon a living animal. I
placed a lively chicken, three weeks
old, in a jar filled with this air : it
grew ſick directly, and was in ſix
minutes near dying, when I took it
out quite motionleſs. It remained
in the open air during ſeveral minutes in a dying condition, after
which it gradually recovered.

I was

I was now thoroughly convinced, that the nitrous teſt failed entirely in ſhewing the degree of ſalubrity of this air ; for it appeared by this method to be nearly dephlogiſticated air, and yet it was ſtill a true poiſonous air*.

I was indeed very ſorry to find this failure in a method ſo well adapted for the exploration of atmoſpheric air. But I am very far from thinking that this exception diminiſhes in any way the real value of the important diſcovery, that *Nitrous air diminiſhes reſpirable air in the proportion to its ſalubrity.* For this teſt holds good in atmoſpheric air, which is the chief object of our experiments.

SEC-

---

* Does this air owe its exploſive nature to the dephlogiſticated air oozing out of the plant ? But this very air becomes alſo exploſive, though it ſtands with a
plant

SECTION XXVI.

Air is one of the moſt changeable ſub-
ſtances in nature, and appearing
under very different forms and
qualities from a variety of cauſes.

THE air of our atmoſphere is
ſeldom during a whole day of the
ſame quality. Its degree of whole-
ſomeneſs is perhaps not leſs ſubject
to variations than its weight and its
degree of heat and cold. The ba-
rometer indicates the firſt, and the
thermometer the other. But thoſe
two inſtruments ſeem to have no
relation to the more or leſs purity
of the atmoſphere, or the more or
leſs fitneſs of the air for the uſe of
reſpiration.

plant during the night, when the plants yield but a
very ſmall quantity of bad air. So that it rather appears to be changed by the plant in a kind of ſimple
explofive air, or a true *fulminating air*, the only yet
diſcovered, as far as I know.

The

The invention of an *Eudiometer*,
or of an inftrument or contrivance,
by which the degree of purity
of the common air, or its fitnefs for refpiration, or rather its
wholefomenefs, can be inveftigated
juft as well as its weight, and its degree of heat and cold, is perhaps
one of the moft extraordinary inventions which ever was made.

We owe this important difcovery
to the Rev. Dr. Prieftley. He found
that nitrous air has the fingular property of diminifhing, or of being
diminifhed by, common air in proportion to its goodnefs; or that the
bulk of the two airs joined together
contracts itfelf in a fo much the
narrower fpace, as the common air
is better, purer, or more fit for refpiration. It will foon appear to
what a confiderable degree of accuracy

racy the Abbé Fontana has brought this truly great difcovery.

We have now in our hands the means of judging, not only of the degree of goodnefs of the common air upon the fpot, but we may with as much eafe alfo judge of the qua- lity of the air of any country, by fending the air of it in clofe bottles. But as the air upon the fame fpot undergoes itfelf continual changes, we can but very feldom expect an accurate agreement of two experi- ments, unlefs made at the fame time, or unlefs a quantity of the fame air be fhut up in a bottle fuf- ficient for different experiments.

Until accurate inftruments fit for fuch purpofes are generally known, and employed with all the attention required, we fhall not be able to judge of that degree of goodnefs which

which the air poſſeſſes for the moſt
part of the year in a country, and
thus to determine the advantages
which would arrive to our conſtitution, in ſpending our lives in one
country rather than in another, on
purpoſe to preſerve a good ſtate of
health, to cure particular diſeaſes
which require a pure air, or to protract our exiſtence in this world in
particular bodily diſpoſitions. We
muſt as yet content ourſelves with
the amuſement of the experiment.

The continual changes which I
obſerved in the atmoſphere daily,
by trying its conſtitution, convinced
me of the too precipitate judgment
of ſome philoſophers, who, though
furniſhed with but indifferent inſtruments, have begun already to
aſſert the degree of goodneſs of
certain places, by one or two obſervations

vations made in the time they paſſed through ſuch a place. But I muſt leave the diſcuſſion of this matter to my reſpectable friend Abbé Fontana, who, in my opinion, has caſt a great light upon this important ſubject; and intends ſoon to publiſh his obſervations on this head. I will add only ſome further reflections upon the changeableneſs of air, its Proteus-like and metaphorical nature.

Since the experiments of the Rev. Dr. *Hales*, we know that air enters the compoſition of bodies, and even ſerves as a kind of cement for the ſtronger coheſion of the conſtituent particles of a ſolid body. By this it ſeems that air may become itſelf a ſolid body, as it conſtitutes ſuch a conſiderable part of ſome particular bodies, ſuch as are, for inſtance, inftance, vegetables, calcarious ftone, nitre, &c. That a fluid body may become a folid, is nothing extraordinary ; we fee that water becomes as folid as a ftone, and remains fo, in a place fufficiently cold. There are perhaps in the world no fubftances which are by their nature *fluid* : for all fubftances yet found may be, by different operations, principally by a fufficient degree of heat, rendered fluid ; and all fluids may be changed into folid bodies by applying to them a fufficient degree of cold. Mercury itfelf was rendered as malleable as any other metal, by Profeffor *Brown* at St. Peterfburg, by a very great degree of cold.

Since that kind of air is known, which goes now under the name of fixed air, and which *Van Helmont,*

mont, called *Gas Sylvéstre*, it has
been imagined by many, that different vegetables contain almost nothing but fixed air, becaufe they faw
that as foon as they began to fer-
ment they emitted really fixed air.
But if from this we conclude that
this very fame vegetable contained
this fixed air, as fuch, concealed in
its fubftance, and exifting there, as
it were in a concentrated or compreffed ftate, almoft as common air
is in a condenfing engine before the
fermentation began, we may poffibly make an erroneous conclufion;
for it may be that this vegetable did
not contain more fixed air as fuch
than inflammable air; but that a
part of the fubftance of the vegetable has undergone fuch a change
by the action of the fermentation
as to become what is now called

I   fixed

fixed air, but what it was very far from being before the fermentation. That this may be the cafe I was induced to fufpect by the following experiments : I fqueezed the air out of different vegetables, keeping them under water, fuch as malva, potatoe-plant, hyofcyamus, apples, &c. I expected to find the moft part of this air fixed air : but I was much difappointed ; for this air was not diminifhable by fhaking it in water. By examining it in another way, I found that the flame of a wax-taper would grow dim in it, and that it was only fomewhat inferior in quality to common air; for one meafure of this air drawn from an apple, with one of nitrous air, occupied 1.24 ; and that expreffed from the leaves of hyofcyamus occupied 1.25. The air expreffed from malva and

potatoe-

potatoe-plants appeared to be somewhat better. This air is undoubtedly the very air of the vegetable
unaltered. I placed all thofe vegetables feparately near the fire in
water, and by examining the air
difengaged from them I found it
to be of a much worfe quality than
that which I obtained by fqueezing;
and by trying the air extracted from
them by actual ebullition, I found it
to be poifonous, and to extinguifh
flame. The air from an apple obtained by boiling was fo bad, that
one meafure of it with one of nitrous air occupied 1.71. Now thefe
very plants, placed in the fame water in the fun-fhine, yield very fine
dephlogifticated air, and by fermentation they yield fixed air. Is it
therefore not probable, that the
very air contained in the plant in

its

its natural ftate was really an air
approaching in quality to common
air; and that the heat of the ebullition had changed this very air into
phlogifticated air, in the fame way
as the act of fermentation changes
it into fixed air, the light of the fun
into fine dephlogifticated air, the
digeftion in the ftomach and the
inteftines of an animal (a great deal
of the air contained in the inteftines,
and all that from which we eafe
ourfelves by the rectum, is pure
inflammable air) and actual fire applied to it into inflammable air, and
the obfcurity of the night into another kind of truly poifonous air?
Could it be faid with any degree of
probability, that one and the fame
vegetable contains thefe fix kinds
of air, fo different in their nature,
and even contrary to one another?

Is

Is it not more reasonable to say that vegetables contain an air, or by whatever name you will please to call it, which by undergoing different operations changes into different sorts of air?

Whoever therefore says, that such or such substance contains such or such air, because he extracts such air from it by the action of fire, by fermentation, or by any other means, may speak erroneously.

Nitrous acid, or spirit of nitre, yields nothing but nitrous air when it is poured upon mercury, copper, iron, &c.; but, when it is mixed with iron filings in a very diluted state, it gives, by the assistance of a moderate degree of heat, a mixture of different airs, partly fixed, partly common air, and partly phlogisticated air, (which experiment I saw

at

at Abbé Fontana's). When this very acid is joined to fome earthy fubftance, or to a vegetable alkaline falt (with which it conftitutes nitre), it yields by the action of the fire nothing but pure dephlogifticated air, in fuch abundance, that the quantity of it is equal to about eight hundred times the bulk of the nitre, as Abbé Fontana found.

Such-like tranfmutations which air feems to undergo are every where obvious in nature. All bodies upon our earth, or almoft all, undergo continually fome alterations, and at laft deviate entirely from what they were before. The plant which affords us the moft wholefome food is perhaps the next to another which draws out of the fame fpot of ground a poifonous juice. The food by which a viper lives

lives changes within his body into a fubftance which has nothing deleterious in itfelf, but in one place of its body a moft virulent poifon is elaborated from it. The fame juice which the root of a tree pumps from the earth is changed into various fruits, very different in tafte and qualities, if different fort of fruits are grafted upon it. An animal body becomes a manure for plants by corruption ; it changes thus in the fubftance of a vegetable ; this, being burnt, changes into afhes ; which, by the action of the fame fire, and by the addition of fome fand and fome calx of lead, changes into fine tranfparent giafs. Thus what is now a part of our body may become in a fhort time a part of a pot or bottle.

I 4　　　The

The three mineral acids themfelves may poffibly be but one,
and the fame acid modified by fome
particular addition, which time may
difcover, to feparate and thus to
change marine acid into nitrous acid,
&c. Some eminent chymifts have
already afferted this as their opinion. More or lefs phlogifton in
one acid than in another may make
the one quite different in nature
from the other. Common air impregnated with phlogifton makes a
poifonous air; and common air,
deprived of it, makes dephlogifticated air; in the one an animal
dies in a little time; in the other
it lives four or five times longer
than in common air. Vitriolic acid
extracts from iron its phlogifton,
and allows it to impregnate the air
difengaged in the act of folution.

Nitrous

Nitrous acid difengages alfo the phlogifton from the iron, but does not allow it to pafs in the air difengaged from it, fo as to make it inflammable. It feems to keep this phlogifton to itfelf; for it is, after the folution, no more to be found in the diffolved iron, when precipitated in the form of ochre; but the fame fpirit of nitre, when diffolving iron in a very diluted ftate, leaves the moft part of the phlogifton with the metal, and rifes in the form of partly fixed air, common air, and what is called phlogifticated air, as was faid above; and by this method iron may be reduced to the moft impalpable powder, all obedient to the magnet, which is a method of making *Æthiops Martialis* of great importance for medical ufes, and was difcovered by an apothecary of Paris.

Paris. Vitriolic acid extracts from calcareous earth, fixed air, and from some kinds of sparrs an air of a wonderful quality, corroding glass itself, which seems to be almost an incorruptible substance, and reducing it into dust by its contact only; and this air, so active upon glass, is by the first approach of water immediately reduced again into the form of the stone out of which it was extracted.

Considering all what is said before, I incline much to the opinion, that the various kinds of air extracted from the different bodies owe, for a great part, their specific nature to the transmutation which they undergo in the operation by which they are obtained; and that they cannot, at least not all, be said to exist in the body in a contracted state

ſtate with more propriety than that glaſs exiſts actually in our body, becauſe, by the action of fire, our body may be changed in a conſtituent part of that ſubſtance; and that fat exiſts in graſs and other vegetables, becauſe in the organs of an animal feeding upon theſe herbs they are partly changed into fat. Thus, when we feed upon vegetables, we do not in reality take in fixed air, exiſting as ſuch in the ſubſtance of that food, and only let looſe or extricated in our bowels; but it is more probable, that ſuch food, undergoing in our ſtomach and inteſtines a kind of fermentation, yields really fixed air, not extricated, but generated by the act of fermentation.

As we have ſeen now, that common air is far from being an unalterable fluid, only to be changed

by

by the addition of fomething, or
by becoming deprived of fomething
extraneous to its own original fimple nature; we can no more be
furprized to find, that the conftitution of the atmofphere remains feldom a whole day the fame, and that
the degree of falubrity is continually
changing. Indeed, in the courfe of
three months, which I fpent in my
folitary retirement, I fcarce found
the degree of falubrity of the common air juft the fame during two
days.

Thofe who are not yet acquainted
with the accuratenefs of Abbé Fontana's new *Eudiometer*, will be much
inclined to believe, that the appearance of fuch continual variations
is more owing to the imperfection
of the method of exploring the air,
than to the real changes happening
in

In our element: and, indeed, I was
much of that opinion, till Abbé
Fontana convinced me of my error;
for, by keeping a bottle full of air
taken from the atmofphere at the
fame time, the conftitution of it is
explored and accurately regiftered;
and examining fome time afterwards
this very air, clofely fhut up in a
bottle, you find the refult of the
trial to correfpond exactly with the
refult of that which was made at
the time when the air was taken
from the atmofphere, and by no
means conform with the refult of
the trial inftituted with the common air of the day, unlefs it fhould
happen that the conftitution of the
atmofphere was juft the fame at
both times. I take this to be a
demonftrative proof of the excel-
lence of this method, as well as of
the

the erroneous judgement which any body might form of the accurate degree of goodnefs of the air of any given place, by examining it once or twice with nitrous air, principally if the obfervator is not in poffeffion of an accurate inftrument for making fuch an obfervation, or if he has not obferved to the greateft nicety all the manoeuvres in the time of making the experiment.

It would be a difficult tafk to difcover as yet the true caufes of that continual fluctuation in the degree of falubrity of the air in the fame place. But it feems to me not improbable, that this inconftancy is to be attributed in general to the natural changeablenefs of the air itfelf, by which it undergoes continual alterations from a variety of caufes, of which a great
number

number are perhaps not to be
traced by human fagacity; and, indeed, if the air of a vegetable is
from the nature of common air, or
air approaching it, changed into
true poifonous air, by applying only
heat to it, as I have faid already,
and that fome more or lefs light
to which a plant is expofed changes
its natural air into the moft falu-
brious or the moft poifonous air,
may it not be fufpected, with fome
degree of reafon, that a great variety of caufes, which have been
till now overlooked, and which vary
themfelves continually, may bring
on a very material alteration in our
atmofphere, fuch as, for inftance,
heat and cold, drynefs and moifture,
light and obfcurity, which I have
already demonftrated to affect the
operation of vegetables upon the

air,

air, winds blowing from different quarters, and conveying airs of different qualities, from diſtant countries, and many other operations of nature, unnoticed as yet?

Water itſelf, one of the ſimpleſt and the moſt unalterable ſubſtances known, ſeems to be changeable into dephlogiſticated air, or at leaſt to contain ſome things which may be transformed into this air by the influence of the day-light; for the green vegetable ſubſtance, which ſerves as a kind of laboratory, in which this ſalubrious air is produced, is formed from the water itſelf. Abbé Fontana made a great many experiments tending to examine the air extracted from different waters by heat. I was preſent at theſe experiments in the ſummer of 1777, being then at Paris. He extracted

5   from

from water of the Seine, and of the aqueduct of Arcueil, an air better than common air, which was a ftep towards the difcovery of ftill better air from fimple water, by fome other way not yet hit upon. Thefe interefting experiments are printed in the *Journal de Phyfique de l'Abbé Rofier*, May 1779.

SECTION XXVII.

On the nature of the air oozing out of our fkin.

AS our **bodies** perfpire continually a watery liquid, either in an invifible way by what is called infenfible perfpiration, or by way of

K   fweat,

fweat, fo a quantity of air feems to iffue continually from the pores of our fkin. This is eafily to be obferved in a cold or warm bath, in which we may clearly fee whole bubbles of this air rifing upon the fkin, and at laft rife to the top of the water. By plunging the hand and arm even in cold water, we may immediately obferve a large number of thofe bubbles every where : and they are the more apparent when the fkin is thoroughly dry before the part is plunged into the water ; and much more fo when it is plunged precipitately into it.

It is however to be obferved, that all the air contained in thofe bubbles, which appear upon the fkin, when a part of our body thoroughly dry is on a fudden plunged under the

the furface of the water, is not
fuch as really iffues out of the
pores ; for, as our fkin is always
covered with fome unctuous matter
which feems to repel water, the
fudden immerfion does not allow
the water to chace before it all the
air fticking, as it were, to the fkin,
but a good deal of it is left upon
it, and forms partly thefe large
bubbles. This feems to be the
more probable, as particularly thofe
places to which thefe bubbles adhere are found quite dry, if obferved attentively, when the part is
withdrawn out of the water. But
thefe very bubbles are in all probability alfo partly owing to air oozing
out of the fkin; for, if they were
nothing but atmofpheric air, they
would not increafe in fize in cold
water, but become fmaller by con-

denfation :

denfation : now they increafe even
to a very large fize in the coldeft
water, and at laft detach themfelves
from the fkin. A warm bath is
not very proper to obtain the air
oozing out of our fkin. Water
having been warmed has loft a good
deal of the air naturally contained
in it, and thus is very apt to abforb
the air oozing out of the fkin. The
beft water for this purpofe is pump-
water frefh drawn.

If we keep our arm, or any other
part of the body, under water, and
rub off all the air bubbles fticking
to the fkin, we fhall fee in a little
while a great many fmall ones
fucceed. But the eafieft way to
convince one's felf of the continual
oozing out of air from our fkin, is
to rub the fkin with the edge of an
inverted glafs full of water, and
long

long enough to keep a good part of
it above the furface of the water in
the time the brim of it is fliding
under water over the fkin. In this
way one may fee an immenfe number of very minute bubbles rife
continually to the top of the water
in the inverted glafs, and gather
in larger bubbles at the top. By
this method I collected, in a little
time, from my arms, a meafure of
this air, which feemed to be partly
fixed air, as it was fomewhat abforbed by the water; at leaft, I
thought to find the mafs of it always lefs than it was before. This
air put to the nitrous teft was found
far from being good refpirable air;
for one meafure of it, with one of
nitrous air occupied, 1.46.

I took a quantity of air in
like manner from the arms of a

healthy

healthy perfon, 19 years old, and found one meafure of it and one of nitrous air to occupy 1.84; which convinced me that the air evaporating from the fkin of young people is not purer than that emitted from the fkin of people more advanced in years; and that if there fhould be any advantage for old people to fleep in the fame bed with young ones, as fome imagine, it cannot likely be owing to their perfpiring a better and wholefomer air from their fkin. It is a very erroneous opinion, and even tending to do material mifchief, that the air of a room, in which a great number of young people have been fhut up, as in fchools, is become very wholefome for old people to breathe in. I have feen fchool-mafters fo ftrongly prejudiced with this notion, that

6   they

they even would not allow the windows of the fchool to be opened, for fear that the young air, as they called it, of the fchool-boys fhould efcape; thinking that breathing this infectious and truly noxious evaporation would prolong their own life.

As I found that the bubbles appearing upon the fkin, when a part of our body is plunged under water, are fo much the larger as the part is put the more precipitately in the fluid, I could fcarce doubt but the air gathered from thefe large bubbles muft be for a great part atmofpheric air, which could not fo quickly detach itfelf from the fkin by the fuddennefs of the immerfion; and I expected, therefore, that this air would give by the nitrous teft a better appearance than that

K 4   gathered

gathered from the fmall bubbles
fcraped from the fkin by the edge
of a glafs. I gathered from another
young and healthy perfon the air
of the large bubbles found upon
the fkin, by plunging the arm fud-
denly under the water. And I found
it approaching more to the nature
of common air, though a candle
could not have burned in it, nor
an animal breathe in it without
anxiety; for one meafure of it
with one of nitrous air occupied
1.40.

CON-

CONCLUSION.

I HOPE the indulgent reader
will excuse in me a fmall degree of
vanity, in flattering myfelf with having difcovered a law of nature hitherto entirely unknown, and hid
till now behind the fcreen of that
awful darknefs which overcafts our
earth during the time it withdraws
its furface from the direct influence
of that all-reviving fummary, the
fun.

I flatter myfelf alfo to have put
beyond all doubt, that the vegetables have a remarkable fhare in
keeping up the falubrity of our atmofphere, by imbibing thofe feptic,
noxious, and phlogiftic particles,
which were communicated to it by
the

the breathing of fo many animals
which inhabit the furface of the
earth, and by many other caufes;
as well as by pouring down a moft
beneficial fhower of purified or dephlogifticated air, which, diffufing
itfelf through the mafs of common
air, counteracts the general caufes,
tending to contaminate our atmofphere, and to render it unfit for
the ufe of refpiration. I was lucky
enough to difcover that the *vegetation* itfelf of the plants has nothing
to do with the cleanfing our atmofphere; but that this great work is
operated by the influence of the
fun's light, exciting and keeping up
the vital and inteftine motion of
thefe numberlefs fans, which the
moft part of plants difplay at once,
juft at the time when the general
tendency

tendency to corruption is increafed by the increafe of heat.

Though we are too much accuf-tomed to look upon the moft obvi-ous operations of nature with a kind of unconcern and indifference, fuch as, for inftance, the vegetation of plants ; yet we cannot look with fo much indifference upon the final caufes of thofe every where obvious fcenes when we difcover them ; for they do not fo much affect the or-gans of our fight and other external fenfes, as they do our underftand-ing, our reafon, our judgment ; by which only we are fuperior to all other living animals. The confi-deration of final caufes gives us to underftand that this great *univerfe* is not the offspring of chance, not coëval with the beginning of time, or of an eternal origin ; but that it has

has been made by an Omnipotent
Being, who, by giving it exiftence,
has, at the fame time, endowed it
with moft wonderful qualities and
powers, continually in action, and
tending with an aftonifhing harmony to one general end, the pre-
fervation of the whole.

An upright mind, averfe to that
manner of living which induces
many to wifh, rather than really to
believe, that this world is not fu-
perintended by an intelligent Being,
takes delight in finding out thofe
deep defigns, which, by their obvious tendency to promote the pre-
fervation of the whole, infpire him
with that awful reverence we owe
to the Supreme Caufe of every thing,
and fill him with that confoling
expectation, that the only being upon earth capable of true reafon, and
of

of tracing the exiſtence of a God in his wonderful works, and of contemplating him in adoration, may expect not to be entirely annihilated after his body is returned into duſt, out of which it took its origin.

But to come back from this digreſſion to the purpoſe, let us conſider how much the real facts drawn from nature itſelf are concordant with the theory deduced from my experiments. If the leaves of vegetables have really a conſiderable ſhare in cleanſing the atmoſphere, it muſt happen, that the time, when our common air is the pureſt, is the ſummer and the winter; for in the ſummer the plants are in their greateſt vigour; and in the middle of the winter the cauſes of general corruption are the moſt

moſt checked by the cold. Now this is juſt what happens. As ſoon as in the advanced autumn the leaves begin to wither and to fall, and to contribute even ſomewhat themſelves to contaminate the air by their corruption, the degree of purity of the atmoſphere is really leſs than it was during the time of the ſummer; and this atmoſphere does not return to its former good quality till the winter is ſet in, and till the remaining tendency to corruption is checked by the increaſe of cold. In the ſpring, when the ſun begins to promote ſomewhat the general tendency to corruption, without having yet influence enough upon the vegetables to make them diſplay their leaves, the common air begins to be leſs fit for reſpiration, till it returns again to its former

mer purity as foon as the leaves are
produced. And this is what Abbé
Fontana found to be a conftant
fact.

If I had more leifure, I fhould be
inclined to expatiate in a wide and
open field of reflections, which prefent themfelves to my mind, 'and to
draw all the confequences which
feem to flow, as from a fountain-
head, from the already mentioned
obfervations.

Is it not probable, that thofe who
labour under confumptive and afth-
matic complaints, and who find the
greateft relief, and fometimes a perfect cure, by retiring to mild climates, where vegetation is lively,
and begins fooner in the fpring,
fhould go to fuch places where the
conftitution of the air is found by
experience to be during the whole
year

year the beſt? But theſe places will not be known till ſome accurate method of examining the goodneſs of common air be in general uſe.

Is it not ſomewhat probable, that it is unſafe for the health of people to ſleep in rooms having windows towards a ſmall open place crouded with the branches of a large tree, ſo hidden from the influence of the ſun as to receive but ſeldom its rays? I remember to have heard people ſay, that it was unwholeſome to ſit under a wallnut-tree, and that they found themſelves affected by its ſhade. But I looked upon ſuch an apprehenſion as one of thoſe popular or vulgar errors which are propagated from father to ſon. I ſhould now be inclined to think, that an apprehenſion of ſome miſchief might not be entirely ill-grounded,

grounded, when such a tree stands, as is often the case in a narrow yard confined by the surrounding buildings.

It is a general belief in the West Indies, founded upon constant experience, that the mangeneel-tree *Hippomane Mancinella* (Linn. Spec. Plant. 1431) throws out very hurtful exhalations, so as to endanger those people, who, ignorant of the nature of this tree, venture to lay down under it.

The plant called *Lobelia Longiflora*, growing also in the West Indies, spreads such deleterious exhalations from it, that a considerable oppression is felt upon the breast in approaching, at several feet distance, this plant, placed in the corner of a hot-house or room. (See the de-

L   scription

scription of this plant in *Jacquini Hortus Botanicus Vindobonenfis*).

The plant called *Dictamnus Albus,* or *Fraxinella,* which is by no means rare throughout almoft all Europe, when in flower fpreads round about inflammable air, which, by the approach of a candle by night, flafhes as other inflammable air does. We know that an animal breathing in this kind of air lofes its life : fo that if a man was to fleep with his head in the middle of the branches of this plant, he might run a rifk of being killed by it.

May we not afcribe the unwholefomenefs of the air of that immenfe plain in which *Rome* ftands to the want of trees and other vegetables? That very plain was, in ancient times, reputed to be a very wholefome country, when it was well cultivated and

and inhabited. And in our days, being not far from a real defert, it is fo notorious for being unwhole-fome, that the people of the country think it highly dangerous to pafs. a fingle night in it, even in the middle of the fummer; whereas in Tufcany, which is peopled and cultivated to the utmoft, one may fleep the whole fummer in the open air without fearing more injury from it than from the air within the houfe. The Pontine Lake, *Lacus Pontinus*, in the dominions of the pope, in which formerly, when cultivated, were numbers of inhabitants, fupplying Rome with the beft productions of the earth, is at prefent a moft difmal defert, fpreading round about it unwholefome and deadly exhalations, fo that fcarce any living animal can breathe this air with-

L 2 out

out foon lofing its health, and finding its deftruction.

The want of proper cultivation
contributes, perhaps, not a little towards rendering the immenfe plains
of Hungary lefs wholefome than
they would otherwife be. The
country round about Vienna is perhaps likewife in want of a fufficient
number of trees.

PART THE SECOND.

Containing a series of experiments made with leaves, flowers, fruits, stalks, and roots of different plants, on purpose to examine the nature of the air they yield of themselves, and to trace their effects upon common air in different circumstances.

SECTION I.

Introduction.

BEFORE I proceed to give an account of the various experiments I made during the course of

this

this fummer, 1779, I muft firft acquaint the reader, that the method which I generally purfued, in putting the different fpecies of air to the nitrous teft, was the fame which the celebrated Abbé Fontana makes ufe of now, and of which he himfelf has not yet given an account to the publick. As I had no right either to claim the invention of his method, or to anticipate the publication of it without his leave, I have afked his confent on this head. He agreed to my requeft very readily, gave me his notes to confult, and even permitted me to get his inftruments engraved; for which purpofe he allowed me to make ufe of his own drawings.

As he had already fhewn me his method of examining the different kinds of air in regard to their degree

of

of falubrity, or fitnefs for refpiration, when I was with him at Paris
in the beginning of the fummer of
1777, and as I have, fince he re-
joined me in London, 1778, feen a
very great number of the like experiments, I provided myfelf with
the fame inftruments, on purpofe to
imitate his method of examining
air, which I found fo accurate, that,
in ten experiments made one after
the other with the fame kind of
air, the refult differed feldom above
$\frac{1}{500}$; that is to fay, that the remaining bulk of the three meafures of
nitrous air, which he joins one after
another to the two meafures of atmofpheric air, is fo alike in the various
experiments made with the fame
common air, that the difference will
feldom amount to more than $\frac{1}{500}$ of
the whole; which accuracy in ex-

L 4   ploring

ploring the degree of goodnefs of refpirable air furpaffes the exactnefs of judging of the degree of heat and cold by the thermometer of Reaumur.

The Abbé has, fince I faw him at Paris, changed fomewhat his inftruments and method of ufing them, or rather corrected them a little; But they remain ftill materially the fame as they were before.

I muft beg the reader to ftop here, and to caft his eyes upon the copper plate and the explication of the figures before he proceeds farther.

The new *Eudiometer*, or inftrument for finding the accurate degree of falubrity, or fitnefs for refpiration of a given air, confifts of different pieces; two of which are the

the principal and abfolutely neceffary. One is a glafs cylindrical tube,
or the *great meafure*, 18—20 Paris inches long, of an equal bore
throughout its whole extent, whofe
diameter muft be of about $\frac{1}{2}$ inch,
or not much lefs, though it may be
larger. This glafs tube has divifions marked upon it, each of exactly three Paris inches. The infide of this tube ought to be rubbed
with fine emery, to take off the
fmooth furface of the glafs; for,
if the furface of the glafs be not
a little rough, the water will remain
here and there in the form of drops
adhering to the infide of the tube,
when air is let up into it; and thus
fo much of the fpace deftined for
the air is occupied by thefe drops,
which renders the column of air
longer than it would have been if
the

the water had run down equally along the fide of the tube; or at leaſt it renders the column of air uncertain in length: each diviſion of this glaſs tube is ſub-divided into 100 equal parts, which are not ex-preſſed upon the glaſs tube itſelf, but engraved for convenience upon a braſs ſlider or cylinder moving along the glaſs tube. This ſlider muſt be open on both ſides, to ſhew the inſide of the glaſs tube, that the height of the column of water in the tube may be ſeen.

The ſecond neceſſary inſtrument is the *little meaſure*, conſiſting of a glaſs tube of a ſimilar diameter with the great tube, and only three inches long. This ſmall tube muſt alſo be made rough on the inſide with fine emery. This little mea-ſure is fixed in a braſs ſocket, hav-

7   ing

ing a flat flider at the orifice of the tube, which, being pufhed in when the tube and the focket are full of air, cuts off exactly the column of air within the tube, and at once fhuts out that quantity of air which is more than the three inches wanted. All that part of the column of air which is thus cut off by the flat flider, is let out by turning or inverting the whole meafure under the furface of the ~~water. This quantity of~~ air fhut up in this little ~~meafure will~~ be conftantly the fame, whatever change may afterwards happen to the expanfive force, or to the elafticity of the air within this meafure.

Abbé Fontana ufes this *Eudiometer* in the following manner: he firft introduces two meafures of the air

air to be examined, then one meafure of nitrous air : at the moment
the two airs come into contact with
one another, he fhakes the great
tube in the water till both airs are
thoroughly mixed together. This
being done, the tube is put in the
water-trough, in a pofition nearly
vertical, to allow time for the water to run down along the infide of
the tube, and to leave the column
of air free. He then flides the
brafs fcale upon the glafs tube, till
the o, or the mark where the fub-
divifions begin, correfpond with
that fpot where the two columns of
water and air meet. He obferves
with what fub-divifion of the fcale
the next mark upon the glafs tube
above the column of water coincides ; which number he writes
down. He then lets up another

meafure

meafure of nitrous air, fhakes the
tube in the moment the two
airs come into contact, and, after
fome repofe, he moves again the
o of the brafs fcale to the place
where the columns of air and water
meet, and writes down the degree
of the fcale which correfponds with
the next mark of the glafs tube
above the water. After this he lets
up a third meafure of nitrous air,
and, after fhaking and repofing as
before, he marks also the degree of
the fcale correfponding with the
next mark of the glafs tube above
the water, and thus finifhes the
whole operation, if the air examined is common air: for no
more diminution of this air would
happen if more nitrous air was
added, as three meafures of nitrous air are fufficient to faturate

3   fully

fully two meafures of any atmofpheric air; he takes particular
care to perform every experiment
in the fame manner, as well in the
handling of the inftruments, as in
the exact time, even to a moment,
of fhaking the two airs together,
of letting the tube ftand by, before
he examines the number upon the
brafs fcale, &c.

After the whole operation is finifhed, he deducts the number of
the fub-divifions of the whole column of air remaining in the tube
from the number of all the fub-
divifions or parts of both airs which
were let up; and the refult gives
exactly the number of parts or fub-
divifions which were deftroyed:
for inftance, if, after the third meafure of nitrous air being let up, the
next mark of the tube correfponds
with

with the number 8 upon the fcale, and if above this mark are remaining three entire divifions of 100 partitions each of the column of air, the quantity of air exifting in the tube amounts to 308 fub-divifions, which being fubtracted from the 5 meafures of both airs employed, or from the 500 fub-divifions of both airs, the remainder will be 192, which is the exact number of the parts or fub-divifions of the two airs deftroyed.

If the air to be examined is dephlogifticated air, he continues letting up one meafure of nitrous air after another in the manner mentioned, till no more diminution takes place. Six, feven, and fometimes eight meafures of nitrous air are required to faturate two meafures of

dephlo-

dephlogisticated air if it be very pure.

What has been already said of this method of putting different airs to the nitrous test will be sufficient, I hope, to guide the reader in imitating it. But he will find in the result of every trial, made with the same species of air, so much difference, that he would mistrust the whole method if he did not observe every minute circumstance in the course of the whole experiment. It has cost the Abbé some years assiduous labour before he reduced this method to that degree of accuracy which it has now acquired in his hands.

Those who wish to perform this amusing experiment themselves will think it worth their while to look over the following necessary cautions

cautions to be obferved, which I extracted from the manufcript of the author.

He reduces the various fources from which errors may arife in this manner of exploring air to twenty; which are, however, not all of equal importance, and may counterbalance in fome meafure one another, fo that one error may correct another. But fome of thefe are of fuch importance, that by overlooking them it may happen that the beft atmofpheric air fhall appear to be a true poifonous one.

Thofe errors may originate principally either from the great tube or meafure, or from the little meafure.

The errors which may originate from the fmall meafure are feven:

M   I. The

I. The firſt error may be committed by the hand, which, by touching this tube in the time of filling it with air, may expand this air by communicating its heat to it. The reſult of this error may amount to two ſub-diviſions.

II. The ſecond error may be committed alſo by the warmth of the hand in which this meaſure is kept, when it is raiſed till the flat ſlider is on a level with the ſurface of the water, in the moment it is puſhed in to cut the column of air within the meaſure from the air to be ſhut out. This error may alſo amount to two ſub-diviſions.

III. The third error may be committed by not keeping the meaſure, in the moment of ſhutting the ſlider, exactly at the height required ; for, if the water within and without the

the meafure be not on a level, the column of air within the meafure may be more or lefs compreffed. This error may amount to four fub-divifions.

IV. The fourth error may depend upon the infide of this tube not be-ing made rough by emery; for water fettles in the form of drops on the fmooth furface of glafs. Thefe drops adhering here and **there to the infide of this tube** render its capacity greater or fmaller. This error may amount to at leaft three fub-divifions.

V. The fifth error depending on this meafure may be owing to the **difference** of time between the fill-ing **this tube** with air and the fhut-ting the flat flider : for after the air is let up into this tube the water runs down its fide for fome time ;

M 2

time; fo that the longer the interval between filling it with air and
pufhing in the flider is, the more
the infide of this tube is cleared
from water, and thus the more air
it will contain. This error may alfo
amount to three fub divifions.

VI. The fixth error, which may
be committed by the fmall tube or
meafure, is indeed remedied in the
meafure in ufe by our author by
the flat flider; but it remains in
the meafure which is ftill ufed by
other philofophers, which confifts
only in a common phial not provided with fuch a flider. The error refulting from the want of this
flider may amount to ten, and even
more fub-divifions.

VII. The feventh error may depend upon the difference in the diameter of the fmall meafure compared

pared with the diameter of the large one; by which difference it may happen, that the dilation of the air within becomes greater or lefs by warmth, as the fubftance of the glafs be thicker or thinner, and that the capacity of the tube itfelf varies for this reafon. The difference of the refult, however, can be but very fmall from this caufe.

Thus, by computing the number of fub-divifions to which thefe feven errors may amount, we find them to be 25. But, as all the five mea-fures of airs are let up in the large tube one after another, thefe errors may, if they were all committed, amount to five times this number, or to 125 fub-divifions.

The miftakes depending from the great tube, or meafure, may alfo amount to feven heads.

M 3   I. The

I. The firſt ſource of error may depend on the inequality in the diameter of this tube, by which a difference of four ſub-diviſions may eaſily reſult in each partition.

II. The ſecond error may depend on the tube not being made rough on the inſide, from which a difference of ſix ſub-diviſions may happen.

III. The third may be owing to the degree of expanſion of the air communicated by the hand in the time this tube is examined to obſerve the length of the column of air. This difference may amount to four ſub-diviſions.

IV. The fourth may conſiſt in obſerving the height of the column of water within the tube, when the water within is not on a level with the water without; by which an

6

error

error of three fub-divifions may be committed.

V. The fifth error may depend on the difference of time between the letting up each meafure and examining the column. From this head a difference of 10 fub-divifions may enfue.

VI. The fixth may be in determining inaccurately the length of the column of air in the tube, which may amount to five fub-divifions.

VII. The feventh error may depend on the tube being kept in a direction more or lefs perpendicular, which may amount to three fubdivifions. All thofe errors refulting from the great tube make together 35, and amount, in the three meafures of nitrous air let up in one

M 4 experi-

experiment, to the number of 105 fub-divifions.

Befides the fources of errors already mentioned, there may happen fome others from accidental circumftances, which may be principally three.

I. The degree of heat of the common air may change during the time you make your experiment, and may occafion a greater or lefs extenfion of the column of air in the great tube.

II. The weight of the air, or its preffure, may alfo change in the interval of inftituting the experiment.

III. The difference of heat communicated to the tube by the body of the obfervator himfelf in the time he is near it to m..ke the obfervation.

Thefe

Thefe three accidental errors, though fmall in themfelves, yet may amount to fix or more fub-divifions. So that all the errors already enumerated may amount, if they were all committed together, to 260 fub-divifions.

Befides all the above-mentioned fources of errors that may be committed, either by indifferent inftruments, or by want of proper attention, there is one which had always vexed me in former times, and which feemed to me, as well as to many others, almoft incorrigible. This is the inconftancy in the quality of the nitrous air, which is found fometimes much ftronger or weaker than at other times, though the fame method of producing it has been obferved.

Of

Of all the metallic fubftances, mercury feems to be the beft to obtain nitrous air of a conftant quality ; but heat muft be applied, if a large quantity is required in a fhort time. I made ufe, a long while ago, of pin-duft, of which a fmall quantity, put in diluted fpirit of nitre, yields all on a fudden a large quantity of nitrous air of an equal degree of ftrength : but as the folution is very tumultuous, and a great deal of the pin-duft, together with the nitrous acid, is apt to rufh out of the phial, I found it at laft better to ufe common copper. I coil ftrong copper wire, neeled fo as to be flexible, up in fmall curls, and fill the phial with them. Thus the nitrous acid, diluted with five or fix times its quantity of water being poured in it, finds a large

and

and always about an equal furface
of the metal expofed to its action,
and yields in a fhort time a large
quantity of nitrous air very conftant
in quality. Brafs feems to me to
give nitrous air of a much more
inconftant quality. Inftead of a
glafs bottle, I often ufe an elaftic
gum bottle, or *caoutchouc*, and, inftead of a bended glafs tube, I take
one made of the fame elaftic gum.
Such tube is eafily made by coiling
up pieces of caoutchouc bottles in
the form of tubes, and fticking them
together by their extremities. This
wonderful fubftance poffeffes a ftrong
power of attraction for itfelf, fo
that two pieces cut with a fharp inftrument will adhere ftrongly together, if joined before the cut and
fmooth edges have been touched by
the fingers, or before they are foiled

in

in any way. To the extremity of
fuch a tube I adapt a hollow glafs
ftopper of a conical form, fo as to
fit almoft all bottles. A brafs ring
forced over the neck of the gum
bottle preffes its fubftance againft
the glafs ftopper, and prevents the
nitrous air rufhing out.

It is to be obferved, that nitrous
acid will at laft deftroy the elaftic
gum bottles by making its infide
brittle, efpecially if the acid is very
concentrated.

Though good nitrous air may
be obtained by many ways, yet this
air lofes gradually its ftrength, and
in a few days, if in contact with
water, becomes very much weak-
ened; fo that it muft be either made
new almoft every day, or we cannot
be fure of the refult of the experiment.

The

The method of Abbé Fontana in putting the different species of air to the nitrous teſt, cut ſhort to the whole difficulty ariſing from the inconſtancy of ſtrength in the nitrous air. By over-ſaturating the air to be examined with nitrous air, it imports little what ſtrength nitrous air has, though even it had loſt almoſt its whole power of abſorbing common air.

In the method adopted by other philoſophers, by which always a certain proportion of nitrous air is added at once to a certain quantity of the air under examination, the reſult is very uncertain if the nitrous air be not exactly always of the ſame quality. But in the method of Abbé Fontana this article is of no conſequence at all.. The only difference ariſing from weak

nitrous

nitrous air in this method is, that more meafures of it are required before the faturation of the air to be examined is compleated.

The reafon of this will appear obvious, if we confider that it is only the true nitrous air which is capable of diminifhing refpirable airs, and that it performs this diminution in the proportion to its ftrength, fo that weak nitrous air will always diminifh common air in the proportion of its own good or bad quality. Now I will fuppofe that the nitrous air, to be added to the two meafures of common air, is become fo weak, either by ftanding, or by the admiffion of any other air, as to poffefs only half the ftrength of good nitrous air. The confequence will be, that as much again of it will be required to faturate

turate the two meaſures of common air ; and thus, after the ſaturation of the two meaſures of common air is compleated, there will remain in the great meaſure, or tube, a column of air ſo much the longer as the nitrous air employed was the weaker. I will illuſtrate it with an example : let us ſuppoſe, that after the three meaſures of ſtrong nitrous air are let up, and the ſaturation of the two meaſures of the air under examination be compleated, the remaining column of air be found equivalent to three meaſures, and eight ſub-diviſions, or to 308 ſub-diviſions ; this number, ſubtracted from the 500 parts or ſub-diviſions of both airs employed, will give a reſult of 192, which is exactly the quantity of both airs deſtroyed. Let us now again

again fuppofe, that the nitrous air employed was fo weak, that, inftead of three meafures, fix were required before the faturation was fully compleated, and that thus the remaining column of air in the great tube occupies 608, inftead of 308, fubdivifions ; we fhall find that the refult will be juft the fame; that is to fay, that, by fubtracting the 608 parts remaining from the 800 parts of both airs employed in the experiment, there will be found exactly 192 fub-divifions deftroyed; and that thus in both cafes the accurate falubrity of the air is afcertained. If fuch bad nitrous air was only at hand as was juft now fuppofed, it follows, that a longer tube ought to be employed.

This obfervation, which I owe entirely to Abbé Fontana, is in my
opinion

opinion of the utmoſt conſequehce, and throws a great deal of light upon the nature of nitrous air, and upon its wonderful property of deſtroying reſpirable air; and it illuſtrates his ingenious theory of this quality, which, I hope, the author will ſoon publiſh, but which I have no right either to claim or to anticipate.

In conſequence of this obſervation we need not be ſo anxious about the goodneſs of the nitrous acid, nor about the ſtrength of the nitrous air.

We have now already examined eighteen different heads, from which errors or differences in this important experiment may ariſe, which may render the whole reſult of the teſt fallacious and entirely uncertain.

N   There

There are two other great fources of errors that may be committed in the examination of any refpirable air by the nitrous teft, which Abbé Fontana has alfo pointed out and corrected.

The firft of thefe two refides in the act of mixing the two airs. If the nitrous air is let up into the large tube, in which was already the air to be examined, and the tube put by for a while without fhaking it, to give time to the two airs to incorporate with one another; or if the two airs are firft put in a feparate veffel before they are let up into the large tube; there will fcarcely ever be two experiments correfponding with one another; the difference will be fo great, and the refult fo uncertain, that it may amount in one experiment to an error of

fifty

fifty fubdivifions, which, being added to the above-mentioned 256,
will make their number in all 306
fubdivifions ; the difference of time
between the moment of the mixture of both airs and the examination will make alfo a confiderable
difference ; even a few feconds may
make a fenfible difference, if the
examination be intended to be
finifhed in a little while ; but if the
two airs after their mixture are left
to ftand by themfelves a long while,
as for inftance 24 hours, the whole
refult will be uncertain, particularly
if the degree of goodnefs of dephlogifticated air is to be examined.

The remaining fource of error
confifts in letting up at once all the
quantity of nitrous air to be employed in the experiment. The uncertainty of the refult arifing from

this

this head is fo much the greater when dephlogifticated air is under examination.

I have now pointed out twenty fources of errors, which may be committed in the examination of the falubrity of different airs, which indeed it would be impoffible to commit all in one and the fame trials, but of which fome are continually committed in the methods hitherto in ufe, and which render the accuracy of the teft as uncertain as the experiment itfelf is: however, I readily acknowledge, that even an accurate philofopher, provided with the fame inftruments which the Abbé ufes, will find it difficult in the beginning to make two experiments with fuch nicety as not to obtain a difference of fome few fubdivifions; but he will by a little

little practice soon be convinced,
that by this method the degree of
salubrity of any air may be ascer-
tained with as much exactness as
the degree of heat and cold by the
thermometer of Reaumur; for the
difference of the result will scarce
exceed $\frac{1}{75}$ of the two airs employed,
and it would be difficult to judge
upon the thermometer of Reaumur
of $\frac{1}{2}$ of a degree of his scale.

Though I have for the most part
made all my experiments according
to the present method of Mr. Fontana, as I was not in possession of a
tube long enough to put the various
airs to the test, according to the original method of the Rev. Dr. Priestley, with as much accuracy as I
could have wished, yet I made a
great many trials in a way not essen-
tially different from his, by letting

N 3 up

up one meafure of common air with one of nitrous air, and fhaking the tube immediately when the two airs came in contact with one another. In examining the dephlogifticated airs, I joined two meafures of nitrous airs to one of dephlogifticated air, fhaking them in the fame way as in the former cafe; but, as thefe two meafures of nitrous air did not always compleat fully the faturation of this kind of air, I added another meafure of nitrous air to it, and fo on till no farther diminution took place.

I have been careful to exprefs the two different tefts in my experiments chiefly in the firft fection, fo that the reader, who is already initiated in the method of Dr. Prieftley, may find how far the refult of my expe-

3   riments

riments will be concordant with the
refult of his own.

When I had an air worfe than
common air to put to the teft, I
found Dr. Prieftley's method ex-
tremely convenient, by adding one
meafure of nitrous air to one of the
air under examination; but I always took care to fhake the two airs
together at the moment of their
mutual contact, on purpofe to a-
bridge the experiment.

The principal thing to be attended to in putting air to the nitrous
teft, either by the original method
of the Rev. Dr. Prieftley, or by any
other, is to perform with the greateft nicety every circumftance conftantly in the fame manner in the
various experiments; for, if you
allow the two airs to ftand longer
one time than another, if you be-

gin

gin to fhake them both together
fooner or later, if you continue
fhaking them for a longer or fhorter
time in one experiment than in
another, if you cail up the account
by obferving the degree of the
meafure, or of the brafs fcale, fooner
or later, &c. you will find in every
experiment very confiderable variations.

SECTION II.

Experiments shewing in general the degree of goodness or purity of the dephlogisticated air which the leaves of plants yield in the sun-shine.

Exp. I. TWO handfuls of *grass*, the roots being cut off, were put in an inverted jar holding a gallon, filled with pump-water, in the sun between eleven and two o'clock; a great quantity of dephlogisticated air was settled at the inverted bottom of the jar, in which the flame of a wax-taper became very brilliant. By the test of nitrous air, according to the present method of Abbé Fontana, the result was as follows: two measures of it being let up in the glass

Quantity of the two airs destroyed.

glass tube, and one measure of nitrous air joined to it, the mark stood after shaking and reposing at 1.92; a second measure being added, it stood at 1.79; after a third measure a 1.68½; after a fourth measure at 1.87½; after a fifth measure at

415 2.85.

By trying it in the other way familiar to Dr. *Priestley*, the result was as follows: one measure of it, with two measures of nitrous air, occupied 0.88.

By pushing this manner of trial farther, as I do, the result was this: by adding to the two former measures of nitrous air a third one, the mark was 1.83; by adding a

319 fourth, it marked 2.81. Thus it appears that by this last method the quantity of both airs destroyed amounts

amounts to 96 fub-divifions lefs
than by *Abbé Fontana's* method.

2. Two handfuls of leaves of a
willow tree were put in the fame
way in the fun-fhine, between eleven and two o'clock; the dephlogifticated air obtained gave, by the
nitrous teſt of Abbé Fontana, the
following refult:

1.96; 1.83½; 1.71; 1.64; 2.55. 445
By the other teſt it gave the following refult: one meafure of this air
with two of nitrous air occupied
0.85; with three 1.75; with four
2.72. 328

3. Two handfuls of *Lamium
Album* were expofed in the fame
way to the fun-fhine, from ten in
the morning till two in the afternoon. A great quantity of dephlogifticated air was obtained of a fine
quality;

Quan-
tity of
the two
airs def-
troyed. quality; it gave, by Abbé Fontana's test, 1.90 ; 1.73¼ ; 1.53½ ; 1.39;

467   2.33.

One measure of it with two of nitrous air occupied 0 98 ; with

340   three 1.60 ; with four 2.60.

4. Two handfuls of leaves of a *vine* were exposed in the same way in the sun-shine, between eleven and one ; a good deal of dephlogisticated air was obtained, which gave, by the test,

415   1.92 ; 1.79 ; 1.61½ ; 1.87 ; 2.85.

One measure of it with two of nitrous air occupied 0.85 ; with

319   three 1.83 ; with four 2 81.

5. One handful of *Becabunga*, the roots being cut off, was exposed in the same way to the open sun-shine, between twelve and four in the afternoon ; a great quantity of dephlogisticated air was obtained of

5

a

a remarkable fine quality, in which the flame of a wax-taper burned with the moſt beautiful brilliancy. It gave, by Abbé Fontana's teſt, 1.87½; 1.73; 1.54½; 1.37; 2.01; 3.00.   500

One meaſure of it with two of nitrous air occupied 0.94 ; with three 1.37 ; with four 2.33.   367

6. A plant of a moderate ſize of common *thiſtle*, juſt before it flowered, was put in the ſame manner in the ſun-ſhine, from eleven till two o'clock; much dephlogiſticated air was obtained of a pure quality; it gave, by Abbé Fontana's teſt, 1.81; 1.51; 1.36; 1.60; 2.60. 440

One meaſure of it with two of nitrous air occupied 0.63 ; with three 1.67 ; with four 2.79.   321

7. Two handfuls of leaves of French *Beans* were in the ſame way
expoſed

expofed to the fun during fix hours; a great quantity of very fine dephlogifticated air was obtained, in which a flame became very brilliant ; it gave, by Abbé Fontana's teft, 2.02;

504   1.92; 1.89½; 1.85; 2.01; 2 96.

One meafure of it with two of nitrous air occupied 0.90 ; with three

345   1.55.

8. Two little plants of *Teucrium Marum* were expofed in the fame manner, during four hours, to the fun-fhine ; they yielded a great quantity of dephlogifticated air of a fine quality ; it gave, by Abbé Fontana's teft, 1.81; 1.59; 1.37; 1.34;

466   2.34.

One meafure of it with two of nitrous air occupied 0.60 ; with three

344   1.59 ; with four 2.56.

9. Some leaves of *Tobacco* were in the fame manner expofed to the

fun during four hours; a great Quan-
tity of
the two
airs de-
troyed. deal of dephlogisticated air was ob- tained, which gave, by Abbé Fon- tana's test, 2.07; 2.06; 2.05; 2.41; 3.39.   361

10. Some leaves of *Cystus Lada-nifera*, a very fragrant plant, were exposed to the sun in the same way during four hours; a great quan- tity of fine dephlogisticated air was obtained; which gave, by Abbé Fontana's test, 1.89; 1.72; 1.56; 1.92; 2.90.   400

11. Leaves of *Juniperus Virgi-niana*, treated in the same way, yielded a large quantity of fine de- phlogisticated air, whose test was 1.91; 1.75; 1.60; 1.79; 2.79.   421

12. Leaves of *Laurus Camphorata*, the camphire-tree, treated in the same way, yielded a good quantity of very fine dephlogisticated air; which

Quan-
tity of
the two
airs def-
troyed.
548
which gave, by Abbé Fontana's teft, 2.01; 1.90; 1.78; 1.73; 1.76; 2.56; 3.52.

13. Some branches of *Cedar of Lebanon* were expofed in the fame manner, from nine in the morning till two in the afternoon, in the funfhine; a great quantity of dephlogifticated air was obtained; whofe teft, by Abbé Fontana's method, was 1.95; 1.77; 1.64; 1.51; 2.25; 477 3.23.

14. Some branches of *Artemifia Pontica* were treated in the fame way; a great quantity of very fine dephlogifticated air was obtained; whofe teft, by Abbé Fontana's method, was 2.00; 1.95; 1.85; 1.79; 454 2.46; 3.46.

One meafure of it with two of nitrous air occupied 0.92; with three 337 163.

SEC-

SECTION II.

Experiments, fhewing the difference in the purity of the dephlogifticated air, which the leaves of the fame plant give at different times.

15. Leaves of an *apple-tree* in an inverted jar full of water were expofed to the fun-fhine from ten in the morning till five in the afternoon; the teft of the air was, 1.80; 1.58; 1.39; 1.86; 2.79. [Quantity of the two airs deftroyed.] 421

16. Leaves of the fame tree expofed during the fame time in the fun-fhine in a green bottle, 1.82; 1.62; 1.71; 2.64; 3.64. 336

17. The fame leaves, which were employed the day before in experiment 15, were expofed again with frefh pump-water in the fun-

O fhine ;

Quan-
tity of
the two
airs del-
tioved.
fhine; the air obtained was remarkably fine, 1.85; 1.69; 1.54;
556 1.38; 1.58; 2.49; 3.44.

18. Leaves of the fame tree expofed to the open air in a very dark
and cloudy day, from five in the
afternoon till fix, had yielded but a
fmall quantity of air, which was
worfe than common air, 1.84;
164 2.36; 3.36.

19. Leaves of the fame tree expofed in the fun from nine in the
morning till twelve, 1.89; 1.71;
417 1.52; 1.60; 2.60.

20. Air from the leaves of a
willow-tree treated in the fame way
as in experiment 1, expofed in the
fun between twelve and two, 2.00;
457 2.12; 2.19; 2.41; 3.43.

21. Air from the fame tree, gathered in a dark, rainy, and windy
day;

day, proved to be of an inferior
quality, 1.89; 1.71; 1.55; 2.26; 3.26.

374

22. Air from the fame leaves gathered between nine and three in a fine warm fun-fhine, 1.90; 1.72; 1.53; 2.22; 3.22.

378

N. B. I think the reafon of the inferior quality of this air to be, that the leaves were too much crouded in the jar, and that thus too many were deprived from light, being fhaded by the others.

23. Air of the fame leaves col- lected in a fine fun-fhine between twelve and 5 o'clock, 1.90; 1.71; 1.49; 1.53; 2.52.

467

24. Air of the fame leaves ga- thered in a fine fun-fhine between two and five, 1.92$\frac{1}{2}$; 1.80; 1.62; 1.60; 2.40; 3.35.

465

O 2

25. Air

Quan-
tity of
the two
airs def-
troyed.

25. Air of the fame, gathered in a warm fun-fhine between three and five, 1.94; 1.79; 1.63; 2.02; 2.99.

401

SECTION III.

Experiments tending to invefligate at what time of the day plants yield the beft dephlogifticated air.

26. THREE jars, each containing a gallon, were filled with pump-water, and two handfuls of *willow* leaves put in each ; they were all expofed at eleven o' clock near one another in a fair fun-fhine.

The

The air of the firſt jar was exa- amined at half an hour paſt two in the afternoon; its teſt was 2.03; 2.05½; 2.03½; 2.02; 2.54; 3.50. 450

The air of the ſecond jar was examined between four and five o'clock; its teſt was 2.06; 2.07½; 2.06; 2.02; 2.08; 3.03. 497

The air of the third jar was examined between ſix and ſeven; its teſt was 2.12½; 2.16½; 2.14; 2.12; 2.56; 3.50. 550

27. Three jars, of a gallon each, full of pump-water, were expoſed at ten in the morning in the ſun, the weather being agreeably warm; each jar contained two handfuls of the leaves of an *Elm-tree*.

The air of the firſt jar was put to the teſt at two in the afternoon; the reſult was 1.90; 1.81; 1.76; 2.66; 3.66. 337

The

The air of the fecond jar was put to the teft at four; the refult was

377 1.91; 1.77; 1.65; 2.19; 3.23.

The air of the third jar was examined between fix and feven; its teft gave 1.97; 1.93; 1.85; 2.16;

388 3.12.

28. Two jars, of a gallon each, full of pump-water, were expofed in a very agreeable warm day in the fun at ten in the morning; each contained two handfuls of *willow* leaves.

The air of the firft jar was put to the teft at three in the afternoon; the refult was 1.10; 2.09;

515 2.08; 2.02; 2.06; 2.97; 3.85.

The air of the fecond jar was examined at five o'clock; it gave by the teft 2.09; 2.11; 2.07;

535 2.08; 2.29½; 2.78; 3.65.

29. Three

29. Three jars as above were Quan-tity of the two airs des-troyed. expofed to warm fun-fhine between eleven and twelve; in each were put two handfuls of leaves of an. *Elm-tree.*

At three in the afternoon the air of the firft jar was put to the teft; it gave 1.91½; 1.93; 1.81; 2.10; 3.10.   390

The fecond jar was examined at five o'clock, the air gave 1.88; 1.67; 1.67; 2.65; 3.65.   335

The third jar was examined at fix; its air gave 1.97; 1.88; 1.84; 2.57; 3.54.   346

N. B. On the day this experiment was made it was a very warm fun-fhine from eight in the morning till two in the afternoon, when the weather began to be more and more dark; we had at three a thunder ftorm; and after the fky was always

O 4   cloudy,

Quan-
tity of cloudy, though it continued to be
the two
airs def- warm weather.
troyed.

SECTION IV.

Experiments tending to investigate what quantity of dephlogisticated air a certain number of leaves yield.

30. ONE hundred leaves of the *Nasturtium Indicum* were put in a jar holding a gallon, filled with pump-water ; it was exposed to the fun in the ordinary way, as in Exp. 1, between ten and twelve o' clock, when fo much very fine dephlogisticated air was fettled at the inverted bottom of the jar, that it filled

filled a cylindrical jar $4\frac{1}{2}$ inches Quan-tity of long, and $1\frac{3}{4}$ inch diameter; which the two airs def- air gave by the nitrous teft 1.94; troyed. 1.82; 1.67; 1.57; $2.45\frac{1}{2}$; 3.44.   456

31. All the air of thefe hundred leaves being taken away, they were again expofed to the fun till feven o'clock in the evening, when about half the above-mentioned quantity of dephlogifticated air was again found in the jar, which proved to be ftill **better than the former; its** teft was **1.99; 1.87; $1.73\frac{1}{2}$; 1.65;** $1.93\frac{1}{2}$; 2.85; 3.79.   520

32. After feparating again this air from the leaves, I replaced the jar in the open air upon the fame place, and left it till next morning at eleven o'clock, when I collected from the fame leaves nearly the fame quantity as the laft time of very fine dephlogifticated air fettled

at

Quantity of the two airs destroyed.at the top of the jar; it gave by the nitrous teft 1.91½; 1.75; 1.58;

511   1.44; 2.20; 3.15; 3.89.

33. Three leaves of a favoy-cabbage, of a moderate fize, were put in a fimilar jar with pump-water, in a fair fun-fhine, between twelve and two o'clock, when two ounce-meafures of dephlogifticated air were difengaged : which gave by the nitrous teft 1.94; 1.78½; 1.63;

364   2.38; 3.36.

34. A handful of the leaves of *Nafturtium Indicum* yielded, in a few hours on another day in a fine fun-fhine, one ounce-meafure of dephlogifticated air, of the following goodnefs, 1.93; 1.76; 1.56½;

504   1.39; 1.99; 2.96.

35. Seven and twenty fingle leaves of a walnut-tree were in the fame manner expofed to the fun-fhine

ſhine in a fair warm day, from
eleven till five o' clock, when they
had yielded about one ounce-mea-
ſure of good dephlogiſticated air.

SECTION V.

*Experiments tending to inveſtigate
the quality of air yielded by plants
in the night, and by day in dark
or ſhaded places.*

36. TWO handfuls of graſs, the
roots being cut off, were put in an
inverted jar of a gallon, full of
pump-water, and placed in a room
during the night, the jar being
covered ſo that no light could come
at it before I examined it. In the
morning

morning a fmall quantity of air was fettled at the inverted bottom of the jar, in which a candle was extinguifhed directly.

37. At nine o'clock in the evening, Auguft 8, when no leaves would yield any more air bubbles, except potatoe-leaves (which always begin the firft to yield air in the morning, and ceafe to yield them at night the laft of all), I filled different jars with pump-water, and put in each a good deal of leaves of fome or other plant or tree, viz. leaves of *oak*, *lime-tree*, *willow*, *yew-tree* or *taxis*, *apple-tree*, *fage* or *falvia*, *artichoke*, *perficaria urens* or *water pepper*, *potatoe* leaves : they were all kept in a room till next morning, when I examined the air which they had yielded.

Early

Early in the morning I found they all had yielded but an inconfiderable quantity of air.

The air of the oak leaves and walnuts was the worſt of all, it was not diminiſhed by nitrous air; that of the willow and the ſage was near as poiſonous; that of the lime-tree was not quite ſo bad; that of the artichoke was ſomewhat better; that of the potatoe-leaves was the leaſt poiſonous of all; however, in none of them a candle would burn even for a moment. That of the *water-pepper* was ſo poiſonous, that it extinguiſhed the flame, though diluted with five times as much common air; the apple-tree leaves had yielded ſo little air that I could not put it to any teſt.

38. Two handfuls of leaves of *French beans* were put in a jar full of

of water, which was kept inverted the whole night in a room; next morning a fmall quantity of air was obtained, which proved to be of a very poifonous quality; it extinguifhed directly a flame. One meafure of it with one of nitrous air occupied 1.94; fo that an animal abfolutely could not have lived in it during one fingle minute.

SEC-

SECTION VI.

Experiments tending to investigate to what degree plants may affect common air in the night, and by day time in shaded places.

39. SOME plants of grass, with roots and all, were put in a jar of a gallon; the jar was inverted upon a dish, and some water put in the dish to keep the plants alive, and to cut off the communication with the external air; in the morning the air in the jar was altered for the worse, the flame of a wax-taper grew dim in it. One measure of it with one of nitrous air occupied 1.24.

5   40. Two

40. Two handfuls of leaves of *French beans* were put in a jar of a gallon; it was kept inverted upon a diſh, and ſome water poured upon it; next morning I found the air ſo much fouled that a candle could not burn in it. One meaſure of it with one of nitrous air occupied 1.39.

41. After having taken out ſome of the air for trial, I placed the jar with the remaining air and leaves in the ſun from nine till eleven o'clock, when I found the air ſo much mended, that a candle could burn in it, and that one meaſure of it with one of nitrous air occupied 1.12.

After this I replaced it again in the ſun till five in the afternoon, when I found the air ſo much mended as to be equal in goodneſs to common air.

6   42. A

42. A little branch of *Cedar of Lebanon* was placed over night in a room under a cylindrical jar of about two ounce-meafures; next morning the air was much fouled by it. One meafure of it with one of nitrous air occupied 1.45.

43. Three jars inverted were placed upon difhes in the manner as in Exp. 39; under one was placed a **potatoe-plant**; under the fecond a plant of common *nightfhade*; under the third a plant of *hyofcyamus*.

Next morning I found the air of all three very much fouled, but that of the *hyofcyamus* the moft of all. In none of them would a candle burn. One meafure of the air of the potatoe-plant with one of nitrous air occupied 1.59; one meafure of that of the *night-fhade* occupied with

P   one

one of nitrous air 1.77; and that of the *byoſcyamus* 1.83.

✱ 44. I placed a very lively chicken, eight days old, under a jar containing a quart full of the air fouled by the *byoſcyamus* plant in the preceding experiment ; it directly grew ſick, and was ready to expire in leſs than half a minute. I took it immediately out of the jar, and put it directly in another jar full of dephlogiſticated air drawn from the leaves of plants. The creature remained for ſome minutes quite motionleſs, ſhewing no ſigns of life but by drawing languidly its breath; it recovered gradually in this air in ſix or ſeven minutes ſo much that it began to ſtand upon its legs and to jelp now and then with a very weak voice. I then took it out of the dephlogiſticated air and put it

upon

upon my hand in the open air ; as foon as it came out of the laſt jar it feemed to grow worfe, and could ſtand no more : however it recovered gradually again.

45. A fprig of *pepper-mint* was placed under an inverted jar upon a difh, with fome water upon it, to keep the plant alive, and to fhut off the communication with the air in the jar, and the air without. It was placed in a warm day in a room againſt the window towards the fun, from eleven till one o' clock. Another fprig of the fame plant, and of the fame fize, was put under a fimilar jar, and placed upon the floor far from the windows in a room very well lighted, but in which the fun did not fhine at that time. By examining both airs I found the air of the jar which was

placed

placed towards the fun rather better than the common air was at that time. But the air of the other jar was lefs good than the common air; for one meafure of it with one of nitrous air occupied 1.13; whereas one meafure of the common air with one of nitrous air occupied 1.06½.

46. **Two handfuls of leaves** of a *walnut-tree* were put into a gallon jar, filled with pump water, and placed about four feet from the wall of the houfe towards the north, and under the fhade of rafsberry-bufhes, fo that no rays of the fun could reach it. After ftanding there during feven hours in an agreeable warm day, fcarce $\frac{1}{10}$ of air was obtained of what thefe leaves ufed to give in the fun-fhine; and this air was of fuch a bad quality that a candle

candle could not burn in it : whereas two handfuls of walnut-leaves
placed in the fun-fhine had yielded
in the fame time a great quantity
of very pure dephlogifticated air,
whofe teft was 1.69; 1.82; 1.69;
1.54; 2.35; 3.34.   466

47. At the fame time two handfuls of *oak* leaves were in the fame
manner placed under the fame
bufhes : the air of thefe leaves was
remarkably foul, for it was very
little diminifhed by nitrous air.

48. The fame was done with
willow leaves ; the air obtained was
alfo of a bad quality, but better
than that of oak or walnut-leaves.

49. The fame was alfo done
with leaves of *elm-tree* ; the air obtained was very foul, as one meafure
of it with one of nitrous air occupied 1.90.

P 3   50. I

50. I placed over-night in a room about an equal quantity of branches of a lime-tree, walnut-tree, a vine, an oak, and a willow, each in a different jar holding a gallon, without water, putting only fome water in the plate upon which the jars were inverted, to keep the branches alive; the jars were left without being covered, and examined between eight and nine in the morning; the air of all was tried by putting one meafure of it with one of nitrous air immediately one after another in the glafs tube.

The air of the lime-tree occupied - - - - 1.24
the air of the walnut-tree 1.25
——————— vine 1.30
——————— oak 1.26
——————— willow 1.23

51. After

51. After having taken out as much air as was wanted for the trial, the jars were placed in the garden in a fine fun-fhine, and the air of them was again examined between ten and eleven.

The air of the lime-tree occupied - - - - 1.08
the air of the walnut-tree 1.07½
———————— vine 1.05½
———————— oak 1.12½
———————— willow 1.07

After which they were again expofed to the fun till three in the afternoon, and their air examined.

The air of the lime-tree occupied - - - - 1.06
the air of the walnut 1.05
———————— vine 1.05½
———————— oak 1.12½
———————— willow 1.07

So

So that the fame air which the plants had fouled in the night was again; reftored to its former purity, and even by fome of thefe plants rendered better than common air, principally by the walnut-leaves, lime-tree, and vine, the teft of the common air being at the time that one meafure of it with one of nitrous air occupied 1.07.

SEC-

SECTION VII.

Experiments shewing that the damage done to common air by the night is very inconsiderable compared to the benefit it receives in the day.

52. TWO handfuls of *Elm-tree* leaves were put in a jar full of water, as in the former experiment, and exposed to the open air from the 14th of August to the 16th, when the air disengaged from them was examined, and found to be good dephlogisticated air; its test was 1.95; 1.85; 1.77; 2.40; 3.50. 350

Quantity of the two airs destroyed.

53. Some grass, kept in a green bottle from the evening till next day at eleven o'clock, had yielded some air, whose test was 1.80; 1.64; 1.54; 2.31; 3.26. 374

54. A

Quantity of the two airs destroyed. 54. A handful of *Perficaria urens*, water pepper, was kept in a green bottle, and expofed to the open air from the evening till next day at ten o'clock, when the air collected from it was found of the following 226 teft, 1.90 ; 1.86 ; 2.78 ; 3.74.

55. Two handfuls of leaves of *Lauro-cerafus* were put in a jar full of water, and expofed to the open air during 24 hours ; a good quantity of dephlogifticated air was obtained, whofe teft was 1.78 ; 1.61 ; 306 2.02 : 2.97 ; 3.94.

SEC-

SECTION VIII.

Experiments shewing that plants have a remarkable power to correct bad air in the day.

56. A sprig of *pepper-mint* put in a jar full of air fouled by breathing (so as to extinguish a candle), and exposed to the sun, had corrected this air in three hours so far that a candle could burn in it. ^{Quantity of the two airs destroyed.}

57. A sprig of a *nettle* was put in a jar full of air fouled by breathing so as to extinguish a candle ; it was placed in a room during the whole night ; next morning the air was found as bad as before. The jar was put at nine in the morning in the sun-shine ; in the space of

two

Quantity of the two airs destroyed. two hours the air was so much corrected, that it was found to be nearly as good as common air.

58. A sprig of *Persicaria urens* was put in a phial full of air fouled by breathing, so as not to allow a candle to burn in it; it was exposed to the sun during an hour and an half, in which time the air was so much corrected that a candle could burn in it. — The same effect was obtained from a sprig of a vine, and that of a camomile plant, and from some rushes.

59. A *mustard* plant was put in a jar; the stem was cut off on a level with the orifice of the jar; the jar was then inverted in an earthen pan containing some water to keep the plant alive, and placed over-night in a room; next morning the air of the jar was found much fouled,

so

fo as to extinguifh the flame of a wax-taper; its teft was, by Abbé Fontana's method, 1.98; 2.87; 3.83. 117

60. It was then expofed to the fun, and examined again after a quarter of an hour, and found already fomewhat corrected; for its teft was then 1.97; 2.84; 3.79. 121

The jar was again put in the open air, when, after ftanding 1½ hour in the fun, the air was found to be remarkably corrected; for now its teft was 2.01; 2.25; 3.24. 176

The jar was again replaced in the fun; when it had been expofed during three complete hours, the air was fo much improved as to be better than the common air at the time; for its teft was 1.95; 2.21½; 2.20 180

The

Quan-
tity of
the two
airs def-
troyeu.

173½

The teſt of the common air was
at that time, by Abbé Fontana's
method, 1.96; 2.25; 3.26½.

See alſo Experiments 40 and 41.

SECTION IX.

*Experiments ſhewing that acrid,
ſtinking, and poiſonous plants yield
by day-time as good dephlogiſticated
air as any others.*

61 A plant of *Hyoſcyamus* was
expoſed in the ordinary way in a
jar full of water in the ſun-ſhine,
from twelve till five o'clock; a great
deal of fine dephlogiſticated air was
obtained, in which the flame of a
wax-taper was very brilliant. One

7   meaſure

meaſure of it with two of nitrous _{Quan-} air occupied 0.93; with three 1.70.

(margin note: Quan- tity of the two airs def- troyed.)

62. Leaves of *Lauro-ceraſus* (one of the moſt active poiſons that exiſts in the vegetable kingdom, when in a concentrated ſtate, killing an animal almoſt in an inſtant, and equally poiſonous when taken in the ſtomach, as when applied to a wound, which laſt effect was lately diſcovered by Abbé Fontana) treated in the ordinary way yield a good deal of dephlogiſticated air. Two handfuls of them expoſed in water to the ſun, between eleven in the morning and five in the afternoon, yielded a good deal of dephlogiſti- cated air, whoſe teſt was 1.87: 1.67; 1.50; 2.04; 3.04. 394

63. Two handfuls of common *night-ſhade*, a ſuſpected plant, ex- poſed

Quan-
tity of
the two
airs def-
troyed.posed in the sun between two and five o'clock, had yielded a great deal of dephlogisticated air, whose test was 1.92½; 1.79; 1.65; 1.52; 495 2.08½; 3.05.

64. I got in the same way good dephlogisticated air from *tobacco leaves*, (see Exp. 9); from *Atriplex Vulvaria*, a plant of a very particular offensive smell; *Cicuta Virosa* or *water hemlock*, one of the most dangerous poisons; and *Sabina*.

S E C-

SECTION X.

Experiments shewing that all flowers in general yield a very poisonous air, though in a very small quantity, and are apt to spoil a great quantity of good air by day and by night.

65. TWO handfuls of flowers of *Marigold* or *Calendula* were exposed in pump-water to the open air during 48 hours; a small quantity of air was obtained, which extinguished flame directly, and was scarcely diminished at all by nitrous air.

66. Two handfuls of *Camomile* flowers were put in a quart jar filled with pump-water, and inverted:

Q   ed :

ed : after two days ftanding in the garden, a fmall quantity of air was obtained, in which a flame was immediately extinguifhed.

67. Forty-five flowers of *Marigold* were put in a quart jar without water, and kept the whole night in a room, the jar being inverted; next morning I found the air fo much fouled by them that a candle would not burn in it. One meafure of it with one of nitrous air occupied 1.43.

68. I placed the fame flowers with the remaining air in the funfhine from nine till twelve, when I found the air ftill more infeƈted. One meafure of it with one of nitrous air occupied 1.54.

69. A few flowers of *Honey-fuckles, Caprifolium,* were placed under a jar of about a pint; when they had

had ftood about three hours in the room, the air of the jar was fo much infected that a candle could not burn in it.

The fame effect was obtained with a fimilar quantity of the fame flowers expofed during three hours in the fun-fhine.

70. A fimilar quantity of thefe fweet-fcented flowers were kept in a pint jar over-night; when they had fo much fouled the air that an animal muft have died in it, one meafure of it and one of nitrous air occupied 1.68.

All kinds of flowers had nearly the fame effect. All of them fouled the air more or lefs, either in a room or in the open air, as well by day as in the night, equally in fun-fhine as in a fhaded place.

Q 2   SEC-

SECTION XI.

Experiments shewing that roots of plants when kept out of the ground yield, in general, bad air, and spoil common air at all times, some few excepted.

Quantity of the two airs destroyed.

71. THREE handfuls of roots of mustard plants, washed clean, were put in a jar full of water in the ordinary way, and exposed to the sun during six hours; when some air was obtained, which extinguished a candle directly.

72. Two handfuls of roots of common *rushes*, well cleaned from dirt, were in the same manner exposed to the sun during seven hours; when a small quantity of air was

6   obtained,

obtained, in which a candle could not burn.

73. One handful of roots of muſtard plants, cleanly waſhed, was put in a quart jar full of water, and three ounce-meaſures of common air let up : after ſix hours ſtanding in the ſun, the air was found changed for the worſe, for its teſt was, by Abbé Fontana's method, 1.95 ; 2.34 ; 3.37.

162

74. A handful of roots of *Becabunga* was expoſed to the ſun in a quart jar filled with water during ſix hours ; a moderate quantity of air was obtained, which, by the nitrous teſt, proved to be as good as common air.

All other roots which I tried yielded bad air, and ſpoiled ordinary air at all times.

Q 3

SEC-

SECTION XII.

Experiments shewing that all fruits in general yield bad air, and infect ordinary air at all times, but principally in a dark place, and in the night.

75. Six *peaches* of a small size were put under an inverted quart jar placed upon a dish, in a room not very light, between two in the afternoon and seven in the evening; when I found the air in the jar so much spoiled, that a candle could not burn, nor an animal live in it. One measure of it with one of nitrous air occupied 1.86.

76. Two of these *peaches*, being put under the same quart jar during

two

two hours, had so much altered the air that a wax-taper could scarcely burn in it a moment, but was ready to go out.

77. Six *peaches* of a small size were placed under an inverted quart jar in the sun between nine and eleven; when, by examining the air, I found it to extinguish a candle. One measure of it with one of nitrous air occupied 1.55.

78. One *lemon* placed under a jar, containing three-quarters of a pint, infected the air so much in a few hours, that a candle burned dim in it.

79. One handful of *filberts* were placed under a jar of two pints during the night; I found the air in the jar so much fouled as to extinguish a candle.

Q 4
80. Six

80. Six fmall *Bergamot* pears were put over-night under a jar of two pints; the air was altered fomewhat for the worfe; a flame grew dim in it. One meafure of this air with one of nitrous air occupied 1.25.

81. Three apples nearly ripe were placed under a two-pint jar over-night; the air was found much infected by them; it extinguifhed a flame. One meafure of it with one of nitrous air occupied 1.48.

82. The remainder of the air in this jar was kept with the apples, and expofed to the fun during feven hours, when the air was become ftill worfe. One meafure of it with one of nitrous air occupied 1.72. A flame was directly extinguifhed in it.

83. Four

83. Four lemons were placed under a quart jar in the fun during feven hours, when the air was changed for the worfe. One meafure of it with one of nitrous air occupied 1.18.

84. A jar holding a gallon was one-third filled with ripe *mulberries*, and expofed to the fun, being inverted upon a difh ; in the fpace of four hours the air in the jar was fo much infected as to extinguifh a flame directly. One meafure of it with one of nitrous air occupied 1.63.

85. *Plumbs* and *blackberries*, ripe and unripe, fpoiled alfo common air in the fun and in the dark.

86. Six apples, as foon as taken from the tree, were directly put in a gallon jar full of pump-water, and expofed to the fun, the jar being inverted

inverted upon a difh; the apples became covered with a great number of fmall air-bubbles. After they had been thus expofed to the fun from ten in the morning till four in the afternoon, a moderate quantity of air was obtained, which proved to be very bad; a flame was directly extinguifhed in it. One meafure of it with one of nitrous air occupied 1.69.

87. Two dozen of young and fmall *French beans* were put in a quart jar full of water, and expofed in the fun from ten till two o' clock: they were covered all over with a great many fmall air-bubbles; the quantity of air collected was but fmall, and in quality fomewhat worfe than common air. One meafure of it with one of nitrous air occupied 1.14; whereas one meafure

fure of common air with one of nitrous air occupied 1.08½.

88. Two dozen of young and fmall French beans were put under an inverted quart jar over-night in a room without water; they ftood till eleven in the morn-ing and were not covered, fo that they had been a long while expofed to all the light of the room. The air in the jar was found fo remark-ably poifoned, that it even furpaffed in foulnefs the air infected by a plant of *Hyofcyamus* (fee Exp. 43); for one meafure of it with one of nitrous air occupied 1.95.

89. I was willing to fee the effect of fuch offenfive air upon a living animal. I placed a very lively chicken eight or nine days old in this air; in the very inftant it en-tered the jar it fhewed figns of the utmoft

utmoſt anxiety, fell down motionleſs, and died in leſs than half a minute. When I ſaw it dying, I took it out with all the expedition poſſible, in order to recover it in another jar full of dephlogiſticated air, which I had kept ready for the purpoſe ; but, notwithſtanding it had not been 20 ſeconds in this foul air, it was quite deprived of life.

Comparing the ſuddenneſs of deſtroying the life of an animal with this air, with that of inflammable air drawn from metals by vitriolic or marine acid, I found that the air fouled by theſe beans was as deſtructive to animal life as the inflammable air itſelf.

90. I placed ſix of theſe beans over-night in a gallon jar inverted upon a plate, on purpoſe to ſee whether

whether such a small number of them could affect observably such a great body of air. I was astonished to find they had so much affected the air, that it was rendered quite unwholesome for breathing; it extinguished a flame; and one measure of it with one of nitrous air occupied 1.34.

91. Three small unripe walnuts were put under a jar of about three ounces measures; from twelve till two o'clock, in a room by day, when the air in the jar was so much spoiled as to extinguish flame. Its test was, that one measure of it with one of nitrous air occupied 1.54.

SECTION XIII.

Experiments shewing that no part of plants improve ordinary air, or yield dephlogisticated air, but the leaves and the green stalks.

Quantity of the two airs destroyed.

92. THE former experiments with flowers, roots, and fruits, are already above related. There remain only the green stalks or branches, not yet covered with the rough skin or bark, and the wood itself, to be examined.

I put some green stalks of a willow-tree, the leaves being stripped off, in a gallon jar filled with pump-water; the jar was exposed, inverted, as ordinary, upon a wall in a warm sun-shine during four hours.

7   They

They became moſt beautifully co-^{Quan-}
vered with an infinite number of ^{the two}
round air-bubbles. A great deal of ^{troyed.}
dephlogiſticated air was obtained,
which gave, by the nitrous teſt,
1.96; 1.87; 1.83½; 2.68; 3.64.   336

93. Some branches of a mul-
berry-tree, covered with grey bark,
were put in a gallon jar full of
pump-water, and expoſed to the
fun. A moderate quantity of air
was obtained, which, being put to
the nitrous teſt, proved to be about
the ſame quality with common air;
its teſt was 2.01; 2.10; 3.10.   190

SEC-

SECTION XIV.

Experiments shewing what kind of water obstructs least the natural operation of leaves yielding dephlogisticated air.

94. AN equal number of willow leaves were put in four different jars, each holding a gallon ; one jar was filled with stagnating water taken out of a pond rather unclean ; the second jar was filled with rain-water collected the day before ; the third with river-water ; and the fourth with water taken fresh from the pump. They were all placed, at eleven o' clock, upon a wall in the fun-shine ; and the air yielded by

the

the leaves was taken out of the jars
at three in the afternoon.

The refult was, that the leaves
put in pond-water had yielded the
leaft quantity of air, and that of no
better quality than common air.
Thofe in the rain-water had yielded
more air, and of a better quality.
Thofe in the river-water had yielded ftill more and better. The pump-
water had yielded the moft and beft
of all.

To be able to judge the better
of the exact degree of purity of
thofe airs, I put them all to the
nitrous teft; the refult was as follows:

Air from the leaves in ftagnating-
water 2.04; 2.20; 3.22. 178

Air from the leaves in rain-water 1.94; 1.96½; 2.69; 3.69. 231

R Air

Air from the leaves in river-water
256 2.05 ; 2.04 ; 2.47 ; 3.44.

Air from the leaves in pump-water 1.96 ; 1.85 ; 1.72 ; 1.64 ;
456 2.47 ; 3.44.

95. I put a handful of leaves of a willow in a jar full of newly diftilled water, and expofed it to the fun during four hours; the leaves gathered fome bubbles upon the under fide, but very few upon the upper fide ; and very little air was obtained, fcarcely enough to put it to the teft; and of this air about $\frac{1}{5}$ was common air, which had flipt in by inverting the jar. It was very far from being dephlogifticated air; it was even worfe than common air.

96. I obtained fome water which was diftilled fome months ago, and put fome leaves of a vine in it. A
fmall

fmall number of bubbles fettled upon the under fide of the leaves, but very few upon the upper furface. The jar was placed in the open air during about five hours, the weather being cloudy. A fmall quantity of air was obtained, which was worfe than common air.

97. I impregnated fome water drawn out of a well with fixed air by Dr. Noot's contrivance, or by the glafs apparatus fold at Mr. Parker's. I put fome leaves of a vine in a jar full of this water ; as foon as they were under water, they were all covered moft beautifully with bubbles. After ftanding about five hours in the garden in a cloudy day, fome air was obtained, which proved alfo, by the nitrous teft, to be worfe than common air, the greateft part of it being abforbed

R 2

by

by the water before it was put to
the teft.

98. I impregnated a jar full of
water with fixed air by means of
falt of tartar and fpirit of vitriol,
according to the method of Dr.
Hulme. I put fome leaves of a
vine in this water, which I found
covered with air-bubbles as foon as
they were plunged under the water,
firft at the under furface, and foon
after at the upper furface alfo.
After ftanding about four hours in
a warm fun-fhine, I found a very
large quantity of air collected at
the inverted bottom of the jar,
which I found by far the greateft
part to be fixed air, as it was abforbed in the water by fhaking. I
put to the nitrous teft that part of
it which remained unabforbed,
and

and found it inferior in quality to common air.

It might be found reafonable to think, that thofe numerous air-bubbles, which appear upon the leaves as foon as put under the fur-face of the water impregnated with fixed air, are owing to the fixed air fettling in the form of thefe bub-bles upon the furface of the leaves.

The fudden appearance and in-creafe of thefe bubbles depend great-ly upon the fixed air fettling on the furface of the leaves ; for any other body gets alfo bubbles in fuch wa-ter ; but the vital motion of the leaves acts its part in this fcene ; for thefe bubbles appear firft on the fame furface of the leaves on which they appear in common water. It appeared, by a variety of experi-ments I made on this head, that

R 3   water

water much impregnated with fixed
air difturbs the natural operation of
the leaves in yielding dephlogifticated air, and that the air thus
obtained was chiefly the fixed air
from the water, and fome little
quantity of air, which is fometimes
better than common air, but for
the moft part much worfe.

SEC-

SECTION XV.

Experiments shewing to what degree of purity dephlogisticated air may be elaborated by vegetables.

IT has appeared in the course of several hundred experiments which I made in my retirement, that leaves of plants in general yield the fineſt air when they are not much crouded together, ſo that the moſt part of them receive the direct influence of the ſun principally in the afternoon between mid-day and ſix o'clock in the middle of the ſummer.

99. I obtained from ſeveral plants ſuch a pure dephlogiſticated air, that the flame of a wax-taper not

R 4   only

Quan-
tity of
the two
airs def-
troyed. only burned in it with fuch a degree of brightnefs that it dazzled
my eyes, but it excited a crackling
hiffing which accompanies the flame
when plunged in pure dephlogifticated air. Among the plants which
yielded the pureft airs were fome
aquatic plants and the turpentine-
trees, from which I always got air
of an eminent degree of purity, fo
that fometimes fix meafures of nitrous air were required before the
faturation of the two meafures of
the dephlogifticated air could be obtained, and that above $\frac{500}{800}$ of the
bulk of the two airs were deftroyed.

100. In September I got from
young leaves of a vine fuch pure
air that its teft gave the following
refult : 1.97; 1.87½; 1.78; 1.68;
470 2.33; 3.30.

101. And

101. And the fame day, from full-grown leaves of a vine ſtill purer, it gave the following reſult: 1.95; 1.85; 1.72; 1.60; 1.61; 2.53.

547

102. The air obtained from the green matter ſurpaſſed in purity the dephlogiſticated air obtained from leaves; this purity was ſo great, that this dephlogiſticated air required eight meaſures of nitrous air to ſaturate two meaſures of it, and that $\frac{645}{1000}$ of the bulk of the two airs were deſtroyed. The dephlogiſticated air, which I obtained from the green matter collected from a ſtone-trough kept full of water near a ſpring upon the high-road, was ſo great, that $\frac{652}{1000}$ of the bulk of the two airs were deſtroyed before the complete ſaturation was obtained.

As

Quantity of the two airs destroyed.

As this green matter is probably of the vegetable kind, I make no doubt but as good dephlogifticated air might be obtained from leaves of plants by fome way or other which I have not yet been lucky enough to hit upon.

However pure this dephlogifticated air may be, that which may be extracted from certain fubftances which do not belong to the vegetable kingdom is ftill fuperior to it, as is the air obtained from nitre and red precipitate.

To give an exact account of the nature of thefe airs, I will place here the refult of the firft mentioned dephlogifticated air, drawn from the green matter, produced by itfelf in the jar. The method of trying it was that of Abbé Fontana, 2.05; 2.01; 1.93; 1.81½; 1.72½; 1.70⅓; 645 2.62½; 3.55.

The

The refult of the teft of the de-Quan-
tity of
the two
airs def-
troyed. phlogifticated air, obtained from the green matter gathered from the ftone-trough on the high-road, 2.08; 1.07; 2.01; 1.92; 1.89; 1.78; 2.54; 3.48.   652

How near the purity of this air approaches to that of the dephlo-gifticated air, extracted by fire from red precipitate, may be feen in the following teft of it: 1.63; 1.28; 93; 59; 27; 58; 1.02$\frac{1}{2}$; 2.50.   750
So that the two meafures of this dephlogifticated air had been re-duced to about $\frac{1}{7}$, and that $\frac{750}{1000}$ of both airs had been deftroyed to complete the faturation.

S E C-

SECTION XVI.

Experiments shewing the effect of plants upon inflammable air.

103. TWO ounce-measures of inflammable air (which was so strong as not to be diminished at all by nitrous air) were let up in a quart jar containing one handful of pepper-mint sprigs; it stood over-night within the house: next day I found the bulk of the air somewhat increased, but still so bad as not to be diminished at all by nitrous air.

104. The same evening I put two ounce-measures of inflammable air in a similar jar with one handful of walnut-leaves; next day I found

I found the bulk of air increased to about $\frac{1}{10}$. One measure of it with one of nitrous air occupied 1.90.

105. Two ounce-measures of inflammable air was also let up in a similar jar with one handful of *Persicaria urens* or water-pepper; next day I found the bulk of the air diminished about $\frac{1}{25}$. One measure of it with one of nitrous air occupied 1.97.

N. B. All these three jars stood in the house from the evening till between twelve and one next day, so that the light of the day may have extricated some air from the pepper-mint and the walnut-leaves.

As neither of these plants could be said to have really corrected this poisonous air, I was curious to see

what

what effect they would have upon
the fame air in the fun.

106. For this purpofe I let up
again two ounce-meafures of the
fame inflammable air in the jar
containing the walnut-leaves Exp.
104; and placed it in the fun between two and five o'clock; when
I found the bulk of the air increafed
to ½, but very little corrected, for
one meafure of it with one of nitrous air occupied 1.89. See, in
Exp. 107, the reafon why thefe
leaves failed to correct this air.

107. I alfo let up two ounce-meafures of the fame inflammable air in
the jar containing the *Perficaria urens*
Exp. 105; and placed it in the fun
between two and fix; when I found
the bulk of the air increafed $\frac{1}{12}$, and
fo much mended, that one meafure
of

of it with one of nitrous air occupied 1.33.

108. I had also let up, in the
same jar in which the pepper-mint
had been the whole night, two
measures of inflammable air, and
kept it in the sun about three hours;
but, having forgot to copy the result of this experiment in my notes,
I repeated it next day by itself. I
let up two ounce-measures of inflammable air in a quart jar, in
which I had put four sprigs of pepper-mint, so that the whole made
up about one handful; I placed it
in a fine sun-shine from one till
half past four; when I examined
the air, I found it increased about
$\frac{1}{10}$, and very much mended; for
one measure of it with one of nitrous air oocupied 1.21; so that it
approached very much to the nature

6   of

of refpirable air ; it exploded however with a loud report.

109. As, in experiment 106, the
two meafures of inflammable air
let up in the jar containing the
walnut-leaves were fcarce corrected
at all in the day time, whereas the
other plants had corrected this air
in a very great meafure ; I fufpected that the walnut-leaves had
fuffered from the inflammable air
in the night time, and that, perhaps,
they had loft their natural power
of correcting this kind of air, or
that fome miftake had been committed ; I thought it therefore ad-
vifeable to repeat the experiment
another day, which I did. Having let up in a jar filled with
pump-water, in which a handful
of walnut-leaves were, two meafures of inflammable air, and left
the

the jar in the fun-fhine from twelve
till five o'clock, I found the air
much corrected, for one meafure
of it with one of nitrous air occupied 1.30. The air was very explofive.

I was now fatisfied that all plants
poffefs the power of correcting inflammable air ; but I wanted to fee
whether plants could reduce inflammable air to the purity of common
air, by letting the inflammable air
remain during two or more days
with the plant.

110. A meafure of inflammable
air was let up in a jar containing
a handful of *Perficaria urens*, and
another meafure in a jar containing a handful of leaves of a walnut-tree. They ftood 48 hours in
the open air, when I examined
them.

S   The

Quantity of the two airs destroyed. The inflammable air put with the walnut-leaves feemed to be corrected fo much as to appear, by the nitrous teft, better than the common air was at the time; for one meafure of it with one of nitrous air occupied 1.03; whereas one meafure of common air with one of nitrous air occupied 1.05. This inflammable air gave the following refult, by Abbé Fontana's 184 teft, 1.91; 2.16½; 3.16.

Having filled a cylindrical jar with this air, I found it explode with an uncommon loud report, which furprized me not a little, and gave me fome apprehenfion that the nitrous teft might fail in fome inftances. The inflammable air let up in the jar with the *Perficaria urens* gave the following teft: one meafure of it with one of nitrous

nitrous air occupied 0.95 ; and with two meafures of nitrous air 1.92. By Abbé Fontana's method it gave 1.90 ; 1.96 ; 2.95.

Quantity of the two airs deftroyed.

205

Thus this air feemed to furpafs far the goodnefs of common air.

111. I then tried it by the flame of a candle, and found it to explode with a very loud report. As I thought the refult of thefe trials very extraordinary, and to afford a remarkable exception in the application of nitrous air to the teft of any air, I repeated each of thefe experiments twice, and obtained conftantly the fame refult.

112. I was refolved, however, to repeat again the experiment : Some plants of *Perficaria urens* were put in a gallon jar, and a good quantity of pure air was let up in the jar. It was kept in the open air from Sunday till Friday follow-

ing,

ing, when it was examined, and
found to be fo poifonous that a
chicken, three weeks old, died in
it in lefs than a minute. It proved
alfo very bad by the nitrous teft;
for one meafure of it with one of
nitrous air occupied 1.80; and the
refult of Abbé Fontana's method
was 2.58; 3.58.

This refult, being quite different
from the refult in Experiments 108,
109, 110, and 111, reftored my
hope that fome blunder had been
committed in the experiments juft
mentioned; I refolved therefore,
if poffible, to difcover this myftery.

113. Two pints of ftrong inflammable air (which could not be
diminifhed by nitrous air) were let
up in a gallon jar containing fome
plants, with roots and all, of *Perficaria urens*, which was placed in

3   the

the garden. After it had ftood 24 hours, the air was examined, and found much mended; for one meafure of it with one of nitrous air occupied 1.23; it exploded with a loud report. It was again re-placed in the garden, and examined after it had ftood 48 hours, when it was found, by the nitrous teft, at one o'clock in the afternoon, near as good as common air; for one meafure of it with one of nitrous air occupied 1.11½. It gave by Abbé Fontana's teft, 2.04; 2.33½; 3.32. And yet it ftill exploded as 168 before.

After this trial it was again placed in the open air, and re-examined the fame day between four and five in the afternoon, when the nitrous teft indicated it to be better than common air; for one meafure

of

of it with one of nitrous air occupied 1.06$\frac{1}{2}$; whereas one meafure
of common air and one of nitrous
air occupied at that time 1.08.

114. This refult convinced me
entirely that the nitrous teft really
fails in this kind of air; for though
it gave all the appearance of good
air, yet it exploded with a loud report; and a chicken placed in it
grew immediately fick, and was
ready to expire in fix minutes, when
I took it out quite motionlefs.

115. The remainder of the fame
inflammable air, which had ftood
during fix days with the *Perficaria
urens* in Exp. 112, without being
much changed, was let up in a jar
containing a plant of muftard. After ftanding 24 hours in the garden
I put it to the teft, when I found it
fo much mended, that one meafure
of

of it with one of nitrous air occu- Quantity of the two airs deftroyed. pied 1.02; one meafure of it with two of nitrous air occupied 2.00. The refult of Abbé Fontana's teft was 1.96; 2.13½; 3.12½. 187¼

So that it already furpaffed in appearance the beft common air; it exploded however with a loud report. I placed the jar again in the garden, and examined the air after it had ftood during 48 hours, when I found it to all appearance ftill more improved; for one meafure of it with one of nitrous air occupied 0.96; and with two meafures of nitrous air 1.80. The refult of Abbé Fontana's teft was 1.97; 1.93; 2.72½; 3.66. 235

It ftill exploded with great violence.

I placed the jar again in the garden during four hours longer in

a fair

a fair fun-fhine, when I found the air ftill better by the nitrous teft; for now one meafure of it with one of nitrous air occupied 0.94; and it gave by Abbé Fontana's method 1.96; 1.87½; 2.44; 3.40.

260

It had not, however, loft its explofive force.

116. Some of the pure inflammable air was put in a jar with an inverted plant of ~~Pyfferia,~~ fo that the root was in contact with the air; it ftood during fix days in the garden, when I found only ⅕ of the air remaining, and this was no longer explofive nor inflammable, but a flame only grew dim in it. So that roots of water-plants have a remarkable power of abforbing inflammable air, as I found by feveral other experiments.

117. I

117. I gathered some inflammable air from stagnating water, by stirring up its muddy bottom. This air was so bad, that one measure of it with one of nitrous air occupied 1.98. One measure of this air was let up in a jar containing a sprig of pepper-mint, the root being cut off. It stood from ten in the morning till four in the afternoon in the sunshine; I found it so much corrected, that one measure of it with one of nitrous air occupied 1.60. It burnt as well as before.

118. An equal measure of the same inflammable air from the stagnating water was let up in a jar containing a small plant of *Persicaria urens*, with root and all. After standing the same time as the former, the air was examined, and found more corrected than the other, for

for one meafure of it with one of nitrous air occupied 1.48 ; but it was as inflammable as before.

It feems to me probable, from the above-mentioned experiments, that plants have a power of correcting even the worft of all airs, inflammable air ; but that they require fome days to perform this tranfmutation, and that one and the fame plant does not live long enough in full vigour to finifh the bunefs, if it is fhut up in a narrow fpace with a certain quantity of this air ; and that this air, after having been in a great meafure mended by a plant, returns again to its former poifonous quality, if it remains with the plant after the vital operation of the plant ceafes, which I apprehend was the caufe of the difference
of

of the event in Experiments 110, _{Quan-
tity of
the two
airsdef-
troyed.}
111, 112, 113, 114, 115.

It appears alſo that plants have a power of changing inflammable air into a kind of air which is not to be known by the ordinary nitrous teſt, and which is the only air I know that explodes without the addition of any other air ; ſo that it ſeems to be by itſelf a true *fulminating air* ; for this inflammable air, after the *perſicaria* plants were changed four times during 16 days, gave at laſt the following reſult, 1.81; 1.56; 1.37; 2.27; 3.25.   375

One meaſure of it with one of nitrous air occupied 0.84; with two meaſures of nitrous air 0.98 ; with three 2.00 ; and yet it had not loſt its exploſive quality, though by this diminution of its bulk with nitrous air it indicated to be far bet-

    ter

ter than common air, nay even to be dephlogisticated air.

I make no doubt but the plants had communicated to this air the quality of being diminished by nitrous air, by mixing with it the dephlogisticated air they yield of themselves; which is also the opinion of Abbé Fontana, to whom I communicated the experiment. But I cannot but think that the plants, by their vital powers, had changed this pure inflammable air into *fulminating* or explosive air, as this quality is given to it in one night, or in a dark place in a few hours; though plants yield no dephlogisticated air in the night or in dark places, but scarcely any air at all, and whatever air they yield is phlogisticated air, unfit for supporting flame. It even seems to me not improbable,

improbable, that living plants not only improve good air, or correct bad air, by communicating their dephlogisticated air to it, but also by a peculiar faculty they possess of purifying the circumamblent air, which they may do by taking to themselves the inflammable particles, or by some other faculty they possess. Air fouled by breathing is thus rendered quite pure again in a few hours by a plant growing in it, as is already shewn above.

SEC-

SECTION XVII.

Experiments towards inveſtigating what plants or trees infect the ſurrounding air the leaſt by night.

119. I PLACED in four different gallon jars an equal quantity, as near as I could, of leaves upon their ſtalks of the following trees, *lime-tree, oak, laurocerafus, walnut.* I placed all thoſe jars over-night in a room, each inverted upon a diſh; in each jar was as much water as would preſerve the leaves alive by keeping the ſtalks wet. Next morning I found the air of all the jars contaminated: that of the walnut-leaves was become unfit for breathing, and extinguiſhed flame;

flame; that of the *laurocerasus* was next in foulnefs to the walnut; then followed the lime-tree; the oak had fpoiled the air the leaft of all.

One meafure of the air in which the walnut-leaves had been, with one meafure of nitrous air, occupied 1.53; that of the *laurocerasus* 1.26; that of the lime-tree 1:16; and that of the oak 1.10.

120. I have obferved that the branches of a vine generally infect the air much lefs by night than moft part of other trees. Cabbage among the culinary plants was, of all I tried, the leaft difpofed to contaminate air.

SECTION XVIII.

Experiments shewing that the purest dephlogisticated air, and the greatest quantity, is yielded by full-grown leaves.

121. I PLACED in a jar full of pump-water the extremity of a branch of a vine containing leaves of different ages, from the full-grown to those which begin only to unfold themselves. The bubbles appeared the first in the old leaves; and they broke out gradually upon the next in age; so that they appeared the latest upon the new-formed leaves. The same proportion takes place also in the size of the bubbles, as well as in
the

the quantity of the dephlogifticated
air obtained from them.

122. I placed in a gallon jar filled with **water** fome old or full-grown **leaves of a vine,** and ex-pofed it to the fun from nine in the morning till two in the after-noon, when a great quantity of very pure dephlogifticated air was ob-tained, whofe teft was 1.95; 1.85; 1.72; 1.60; 1.61; 2.53.   547

122. I placed in another jar of the fame fize a fimilar quantity of young leaves of the fame vine, and expofed them to the fun during the fame time. I obtained a good quan-tity of fine dephlogifticated air, but **lefs,** and of an inferior quality, than that obtained from the old leaves. Its teft gave 1.97; 1.87½; 1.78; 1.68; 2.33; 3.30.   470

T   S E C-

SECTION XIX.

*Experiments shewing that the sun by
. itself, without the assistance of
plants, does not improve air, but
renders it rather worse.*

124. TWO jars, half-full of air
taken from the atmosphere at the
same time, and half-full of pump-
water, were left by themselves during four hours, the one exposed to
a bright sun-shine, the other placed
within the house, only two steps
from a door opening in the garden.
The air kept in the house gave,
in six different trials, constantly the
appearance of being better than
that of the jar placed in the sun.
One measure of the air kept within

5   doors

doors with one of nitrous air occupied $1.06\frac{1}{2}$; whereas that expofed
to the fun occupied $1.08\frac{1}{2}$.

I muft however acknowledge, that
this experiment ought to be repeated
more than once, to put the fact out
of any doubt. I made it the very
laft day of my ftay in the country,
and thus had no time to repeat it.

SEC-

SECTION XX.

Experiments tending to investigate the most accurate and expeditious way of putting common air to the test, on purpose to judge of the salubrity of any country.

I HAVE already said enough, in the introduction to the second part of this work, of the accuracy with which this difficult and important investigation may be made by employing the instruments of Abbé Fontana; but as there is much more attention and dexterity required to judge with the greatest nicety of the degree of salubrity of the atmospheric air than of any other, as the other airs are of much less

importance

importance to mankind; I referved
this article for the laft, not difcon-
tinuing to purfue my experimental
enquiries till the book was already
nearly printed off. Befides, it was
but in the middle of September
that I got the brafs tube, exprefled
in Fig. I. in which the glafs tube
or great meafure is fufpended; fo
that the column of water within and
without the glafs tube be at a perfect ~~level, which is~~ neceffary to
obtain a refult conftantly the fame
with the fame air. This brafs tube
is a valuable addition to the apparatus.

I had before that time made experiments every day with the atmofpheric air, placing the glafs tube
in a cylindrical jar filled with water,
and lifting up the glafs tube till
the extremity of the column of

water

water within the tube was on a level
with the brim of the jar; always
taking hold of the glafs tube by
means of a piece of linen folded
five or fix times, and thoroughly
imbibed with water, to prevent the
warmth of my hand communicating
itfelf to the glafs tube.

Though I ftill think that Abbé
Fontana's method of examining atmofpheric air is the moft accurate ;
yet, as I had it more in view to trace
nature in the operation of vegetables than to examine the degree of
falubrity of the common air, I en-
deavoured to abridge this trial as
much as poffible, on purpofe to fave
time. For this reafon, I got at laft
in the habit of making this trial
in the time of a minute or two,
and found a furprizing accuracy in
the refult. This mode is in fome
degree

degree compofed of the methods ufed by the two moft eminent philofophers in this branch of natural knowledge, the Rev. Dr. Prieftley, and Abbé Fontana. It is this: I let up in the little meafure as much common air as will fill it; after which, I take hold of its brafs flider, and keep it under water exactly 15 feconds, when I lift it up till the brafs flider be on a level with the water of the trough, and fhut the flider, to cut off the column of air within the meafure; I then invert the meafure under water, to let out all the air which was remaining under the flider. I let up immediately this meafure of air in the large tube, and fill the little meafure in the fame manner with nitrous air newly made from red copper, in the manner explained p.

T 4   171;

171; which being alfo let up in the large tube, I begin to fhake forcibly this tube in the water-trough exactly 30 feconds (beginning the motion precifely at the moment the two airs come into contact), and place it directly afterwards in the brafs tube, and let it ftand thus in the middle of the trough for the fpace of one minute, pouring continually water upon it, to bring the temperature of the glafs tube to that of the water; for, holding it in the hand while fhaking, it receives fome degree of heat from the hand, and of courfe the column of air within is rarified. I then flide the glafs tube up or down within the brafs tube, which is filled with water, till the two columns of water come to a level with each other, and with the o of the brafs meafure,

meafure, as is expreffed in the plate
by BB in Fig. I. Then I obferve
with what number of the fcale the
firft divifion of the glafs tube above
the column of water coincides,
which fhews me at once how many
fub-divifions are remaining from
the two meafures of airs, or from
the 200 fubdivifions let up in the
tube, and thus indicates the degree
of goodnefs of common air, or indeed of any air approaching in
goodnefs to common air, or being
of an inferior quality. But this
method will not do in examining
dephlogifticated air, as this air re-
quires more nitrous air to bring it
to a full faturation. By this fimple
and eafy method the whole operation is performed in three or four
minutes ; and its accuracy is fuch,
that frequently in ten trials, made
with

with the fame common and nitrous air, the difference of the refult does not amount to $\frac{1}{200}$ of the bulk of both airs.

The different degrees of falubrity will be found in general to lie between 103 and 109; at leaft, I found it almoft always to be within thefe two extremes: that is to fay, that of the bulk of the two airs the remaining column will be found to occupy between 103 and 109 fubdivifions. The magnifier applied to the brafs tube (D, Fig. I.) affifts greatly the accuracy of the obfervation.

This fimple method conftantly fhewed me all the variations in the conftitution of the atmofphere, in regard to its fitnefs for refpiration, which I could difcover by any other method.

A glafs

A glafs tube longer than that which I had at hand would bear a larger fcale, and thus indicate with ftill more accuracy the goodnefs of the air : but the two meafures of air let up muft not fill more than one half of the glafs tube, for otherwife it could not be fhook in the water without danger of fome bubbles of air coming out of it, or rufhing in it, by the force ~~of fhaking it up and down.~~

In the works of the Rev. Dr. Prieftley, one meafure of common air is faid to occupy fometimes 120, and even more fubdivifions, which is owing to his peculiar method. He firft joins the two airs together in a feparate jar, and allows them to ftand a certain time to incorporate one with the other; after which, he lets them up in his large
tube

tube exactly divided, and fees at once, without any fhaking, how much of the two airs is deftroyed. If this method is purfued accurately, and if the fame interval of time is obferved between joining the two airs and letting them up, the refult will, however, be found different in different experiments, as Dr. Prieftley makes no fcruple to allow.

I made a great many experiments to find out the reafon of this difference; but this tafk I leave to Abbé Fontana, who commenced his enquiries on this fubject prior to me. I will only relate one of my own, which will fhew the reader what refult he may expect from his experiments, though performed in the moft regular manner.

I filled

I filled a jar with common air, and put one meafure of it, with one of nitrous air freſhly made, in five veſſels, each, of a different diameter, to incorporate with each other without moving them : after an hour's time, I let up the airs of the different jars into the large tube or meaſure, when I found that the column of air occupied ſo much the greater ſpace as the veſſel in which they ſtand was of a leſs diameter ; but none of theſe airs were diminiſhed near ſo much as they were when ſhook immediately together in the way above-mentioned. It is very remarkable, that I could ſcarcely reduce any of theſe airs afterwards to a leſs bulk, though I ſhook them very forcibly in the large tube after I had examined them.

125. The

125. The common air ufed for this experiment proved to be of fuch a degree of goodnefs, that in fix different trials, made one after another in the expeditious manner explained, the two meafures occupied 1.06½ exactly; whereas the fame common air with the fame nitrous air, after ftanding an hour in the five different veffels, gave the very different refults expreffed in p. 287.

By repeating the fame experiments at different times, the fame refult was obtained as to the difference of the remaining bulk of airs kept ftanding in veffels of different diameters; but there was commonly a difference of fome fubdivifions even in the experiments made with the fame glaffes.

The

The bulk of the two airs kept in the veſſel of the largeſt diameter occupied in the glaſs tube without ſhaking 1.10½, | And after being ſhook when firſt examined in the glaſs tube, 1.10½

That in the glaſs of the next in diameter 1.23½, 1.22

That in the glaſs of the next diameter 1.28½, 1.28

That in the following 1.35, 1.35

That in the glaſs of the ſmalleſt diameter of all 1.44, 1.43

POST-

POSTSCRIPT.

AS I went on with my experiments during the whole time this book was printing, I continued to difcover more and more the fecret operations of nature in regard to cleanfing our atmofphere. I have carefully regiftered in my notes the refult of the experiments, which I may poffibly communicate to the publick in a fecond volume, together with fome more deductions which I may draw from my remarks.

Though I am obliged abruptly to ftop my further refearches, I cannot difmifs the reader without acquainting him, that, as foon as the warm weather began to ceafe, and the autumnal colds to fet in (the thermometer of Fahrenheit being un-

7   der

der 50 in the shade, which had been in the time of the other experiments in general between 70 and 83), the leaves, fruits, and roots had lost a good deal of their mischievous influence upon the circumambient air in the night, and by day in shaded places, though they had lost nothing of their salubrious power in yielding by day dephlogisticated air ; but that the flowers seemed to have lost very little or nothing of their malignant effluvia by which they contaminate the surrounding air; and that water standing by itself, or with plants in it, loses by the sun-shine, or rather by the warmth communicated to it in the sun, the faculty of promoting, or rather of not obstructing, the plants yielding dephlogisticated air ; but that it recovers almost to an

U equal

equal degree its former faculty, by the coldnefs of the night. Water, in which I found ice in the morning, and which the day before obftructed the leaves in yielding a tolerable quantity of bubbles, was fo much recovered, when it was heated by the fun, that frefh leaves put in it yielded air-bubbles very brifkly, when the thermometer plunged in it was at 37.

From what has been faid in the nineteenth Section, as well as from other experiments, I am more and more induced to believe that our atmofpheric air is a fubftance of a very changeable nature, and that it is, in common with a great many other fubftances, equally liable to become worfe, or of undergoing a kind of corruption by the increafe of heat; and that this tendency to

3   corruption

corruption is checked by the vital operation of the plants in the fummer, and by the cold in the winter. By this obfervation, we may perhaps be induced to believe, that thofe countries which are very hot in the fummer, and are little or not cultivated, as is a great part of Hungary and the country round about Rome, are not only expofed to have their air contaminated by the breathing of animals in it, and by the corruption of many other fubftances, but alfo by the corruption which the air itfelf is liable to undergo during the heat of the feafon; and which mifchief can chiefly be remedied by making a fufficient quantity of vegetables grow in them, principally trees. Draining the marfhes, and preventing inundations by keeping the ri-

U 2   vers

vers within their bounds by dykes, and by cutting canals to let out the waters, will greatly affift the operation of vegetables, which would be infufficient to cleanfe the atmofphere of low countries, without this great caufe of corruption, owing to marfhes, being removed.

EXPLA-

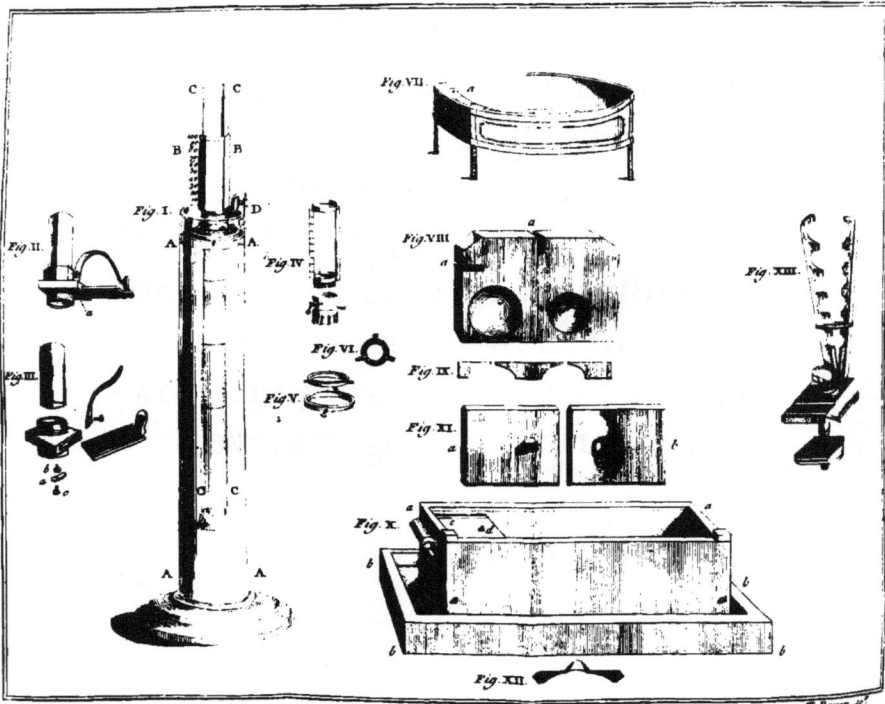

Fig. I. Fig. II. Fig. III. Fig. IV. Fig. V. Fig. VI. Fig. VII. Fig. VIII. Fig. IX. Fig. X. Fig. XI. Fig. XII. Fig. XIII.

T. Bewen sc.

EXPLANATION of the FIGURES.

Fig. 1. THE great meafure, or great glafs tube, in its fituation in the experiment of examining air, with the brafs fcale upon it, and the magnifier ftuck to the brafs tube, on purpofe to adapt accurately the firft mark of the fcale to the beginning of the column of air. The glafs tube is fufpended upon the two brafs rings or gingle (in the way the common fea-compaffes are fufpended) to keep it always in a perpendicular fituation. AAAA is the brafs tube full of water, in which the glafs tube fixed to the brafs meafure is fufpended. This brafs tube is reprefented tranfparent on purpofe to fee in what manner

U 3

the glafs tube is fufpended in it.
B B, the brafs fcale of three Paris
inches divided into an hundred parts,
C C C C, the glafs tube or great
meafure, whofe lower and open extremity is fecured by a brafs ferrule.
D, the magnifier for the more accurate obfervation.

Fig. II. The fmall tube or meafure fixed in its brafs focket. *a*, is
an elaftic piece of brafs, having a
pin paffing through a hole in the
under part of the focket, which
pin is pufhed upwards by the elafticity of the piece *a*, and enters in
a cavity on the under fide of the
flider, made on purpofe to ftop it,
and to prevent the bended fteel
fpring from forcing the flider entirely out of the focket. N.B. I think
that this machine may do very well
without

without the bended fteel fpring,
and therefore I did not put it to
that which I made ufe of.

Fig. III. The fmall tube or meafure, with the brafs focket, flider,
and fprings, all taken afunder to fee
their fhape. The fcrew *c* under
the elaftic piece *a*, is to be fixed to
the under part of the fhoulder of
the brafs focket (at *a*, in Fig. II.),
to fix the elaftic piece *a* to it.
The pin *b* fupported by the piece *a*,
and paffing through a hole in the
under part of the fhoulder of the
focket, enters a cavity made in the
under fide of the brafs flider, when
this flider is drawn out as in Fig. II.
and thus ftops its coming out entirely.

U 4   Fig. IV.

Fig. IV. The brafs fcale with its under-piece to be fcrewed to it, and ferving to embrace clofely, by its fpring, the glafs tube, fo as to be fufpended by it upon the brafs gingle, exprefied in Fig. V. The infide of this under-piece ought to be lined with a piece of fponge, on purpofe to prefs foftly againft the glafs, and to prevent its being fcratched by fliding the glafs tube up and down againft the brafs,

Fig. V. The gingle, or the two brafs rings, fuch as are ufed in common fea-compaffes, whofe move-able axes act contrary to one ano-ther, to give the body fufpended upon them every poffible motion, and thus to keep it in a perpendi-cular line.

N. B.

N. B. I keep my tube fufpended fimply upon a brafs ring, foldered a little way within the brafs tube, which does very well.

Fig. VI. That part of the under-piece of the brafs fcale by which it is fupported upon the gingle.

Fig. VII. The wooden-trough full of water, in which the whole apparatus is ufed. This trough ought to be (in the infide) 2 feet long, 13 inches deep, and 17 inches wide. The board *a*, upon which the jars, &c. are placed, ought to be fixed at the diftance of $3\frac{3}{4}$ inches from the brim; the length of the board ought to be about 9 inches, the thicknefs of it two inches. The trough ought to be kept full
of

of water except about two inches
from the brim.

Fig. VIII. The board (expreſſed
by *a*, in Fig. VII.) by itſelf, and
inverted. It has two funnels hollowed out on its under-part, which
is in this figure repreſented upper-
moſt : the orifices of the funnels
are repreſented by the two round
holes, one of which muſt be larger
than the other. *a a* repreſent two
oblong inciſions to receive the extremity of the bended tubes, through
which the various kinds of air are
let up into the inverted jars placed
upon the board.

Fig. IX. A cut of the two funnels hollowed out in the board, repreſented by Fig. VIII.

The

The reſt of the figures are only intended for thoſe who ſhould like to engage farther in this entertaining branch of natural philoſophy, and to produce thoſe kinds of air which are liable to be altered or abſorbed by water.

Fig. X. A wooden trough to be filled with mercury, for ſuch experiments with air as cannot be done in water. Many kinds of airs are abſorbed themſelves by water, as is fixed air, and all thoſe aërial fluids which ſhould rather be claſſed among vapours, as alkaline air, acid air, &c. of which an account may be ſeen in the works of Dr. Prieſtley, and which will ſoon be treated in a more ample manner by Abbé Fontana.

Air

Air extracted from fpars cannot be examined, nor even obtained, but by making ufe of mercury inftead of water : for this fingular air, which corrodes glafs, is im- mediately reduced into ftone by the firft contact with water.

This trough confifts of two dif- ferent ftrong wooden boxes. *a a* is the box containing the mercury; it is in the infide 11½ Paris inches long, 4 inches and 2 lines deep, and 4 inches and 2 lines wide. The board *c* is placed at one inch and 2 lines diftance from the brim, and is 7 lines thick. The orifice *d* of the funnel hollowed out of the un- der-fide of this board is two lines above the furface of the board. This box is placed within another larger box equally ftrong, *b b b b,*

which

which ferves to receive the mercury fpilt by moving the veffels
in the other box.

Fig. XI. The board (reprefented
by *c* in Fig. X.) of the box containing mercury, reprefented in *a*
as it is fixed in the box; and inverted in *b*, on purpofe to fee the
funnel hollowed out in the under-
fide of it.

Fig. XII. A fection of the board
of the box (reprefented by Fig. X.
c), on purpofe to fhew the form of
the funnel, and the manner of fix-
ing this board in the box, by letting its floping edges in a groove
cut out in the fubftance of the box,
fo that the mercury cannot pufh
it up, but that it may be taken
out at pleafure.

Fig. XIII.

Fig. XIII. A kind of *forceps* or tongs, to receive the necks of different veffels, in which air is to be extricated by heat. It is fixed by the fcrew to the brim of the water-trough, or to the box of mercury; and the neck of the glafs veffel is fqueezed between the two branches by means of the moveable ring, by which they may be more or lefs fqueezed together, according to the fize of the neck of the glafs. This inftrument is very ufeful for different operations, which, without its help, would require an affiftant to hold the glafs to keep it from falling.

INDEX.

I N D E X.

INDEX.

X   air,

INDEX.

air, p. 288. but not of flowers, p. 281.—Does not obstruct the vegetables yielding dephlogisticated air, ibid. —Checks the corruption of common air by itself, p. 290.

Cystus Ladanifera, its air explored, p. 10.

Cultivation, neceffary to keep the atmosphere wholesome proved through the principal part of this work.—Examples to prove this affertion near *Rome,* near the *Lacus Pantinus,* and in *Hungary.* See these words.

D.

Death, to what caufe fudden death fometimes owing, p. 51. 49. 55.

Dephlogisticated air. Its nature, p. lvii.—Is heavier than common air.—Final caufe of it, p. 52.— How to examine it, p. 159.—How to obtain it from plants, p. 14.—In what manner it oozes out of the leaves, p. 17.—Its different degrees of goodnefs, exp. 99—101.—Undergoes a tranfmutation in the plant, p. 25.—That which is yielded by the green vegetable matter examined, exp. 102.—That from red precipitate examined, exp. 102.—New method of procuring it at a cheap rate in any quantity wanted, p. xlv.

Dictamnus Albus. See *Fraxinella.*

Difeafes, new method of curing many, p. xlvii.

E.

Elastic Gum. Its peculiar attraction to itfelf, p. 171.— bottles of this fubftance ufed for producing nitrous and
                                                              inflammable

Helmont,

INDEX.

Y   *Laurocerafus.*

5

Light

Night,

Pontine

The

INDEX.

Spurge.

INDEX.

Spurge. See *Euphorbia.*

Stalks. Green ftalks yield dephlogifticated air, exp. 92.

Strawberry leaves, peculiar manner of yielding dephlogifticated air, p. 19.

Sun. Its light is the productive caufe of the dephlogifticated air ·from plants, p. 28.—By no means its
warmth, p. 21.—Contaminates air by itfelf, without
the affiftance of vegetables, p. 274.

T.

Tobacco yields dephlogifticated air, exp. 9. 64.

Taxis. See *Yew-tree.*

Teft. The nitrous teft for examining air, found out by
Dr. Prieftley. See the words *Prieftley, Nitrous air, Air,
Atmofphere.*—Reafon of the different refults in putting
the fame air to the teft, inveftigated, p. 284. This
difficulty remedied, 281.

Thermometer, at what degree it ftood during the time thefe
experiments were made, p. 288.

Thiftle, its air, exp. 6.

Tranfmutation of air, p. 116.—Of bodies every where to
be found, p. 188.

Trees. What trees to be planted for the wholefomenefs
of a country, p. 93.—Unwholefome, when growing
in a fmall place, p. 144.

Tufcany. Why its air very wholefome, p. 147.

Vegetables.

V.

Water.

INDEX.

W.

INDEX.

Y.

Yew-tree, or *Taxis,* its air by night, exp. 37.

Young Persons. If wholesome for old people to sleep with young, p. 134.

www.ingramcontent.com/pod-product-compliance
Lightning Source LLC
Chambersburg PA
CBHW030901270326
41929CB00008B/518